Biofunctional Textiles and the Skin

Current Problems in Dermatology

Vol. 33

Series Editor

G. Burg *Zürich*

KARGER

Biofunctional Textiles and the Skin

Volume Editors

U.-C. Hipler *Jena*
P. Elsner *Jena*

53 figures, 7 in color, and 29 tables, 2006

KARGER

Basel · Freiburg · Paris · London · New York ·
Bangalore · Bangkok · Singapore · Tokyo · Sydney

Current Problems in Dermatology

Library of Congress Cataloging-in-Publication Data

Biofunctional textiles and the skin / volume editors, U.-C. Hipler, P. Elsner.
p. ; cm. – (Current problems in dermatology ; v. 33)
Includes bibliographical references and index.
ISBN 3-8055-8121-1 (hard cover : alk. paper)
1. Contact dermatitis. 2. Textile fabrics–Physiological aspects. 3. Biomedical materials. I. Hipler, U.-C. (Uta-Christina) II. Elsner, Peter, 1955- III. Series.
[DNLM: 1. Skin Physiology. 2. Anti-Infective Agents, Local–therapeutic use. 3. Biocompatible Materials–therapeutic use. 4. Dermatologic Agents–therapeutic use. 5. Skin Diseases–immunology. 6. Textiles–microbiology. W1 CU804L v.33 2006 / WR 102 B6157 2006]
RL244.B56 2006
616.5′1–dc22

2006010081

Bibliographic Indices. This publication is listed in bibliographic services, including Current Contents® and Index Medicus.

Disclaimer. The statements, options and data contained in this publication are solely those of the individual authors and contributors and not of the publisher and the editor(s). The appearance of advertisements in the book is not a warranty, endorsement, or approval of the products or services advertised or of their effectiveness, quality or safety. The publisher and the editor(s) disclaim responsibility for any injury to persons or property resulting from any ideas, methods, instructions or products referred to in the content or advertisements.

Drug Dosage. The authors and the publisher have exerted every effort to ensure that drug selection and dosage set forth in this text are in accord with current recommendations and practice at the time of publication. However, in view of ongoing research, changes in government regulations, and the constant flow of information relating to drug therapy and drug reactions, the reader is urged to check the package insert for each drug for any change in indications and dosage and for added warnings and precautions. This is particularly important when the recommended agent is a new and/or infrequently employed drug.

www.karger.com
Printed in Switzerland on acid-free paper by Reinhardt Druck, Basel
ISSN 1421–5721
ISBN-10: 3–8055–8121–1
ISBN-13: 978–3–8055–8121–9

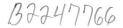

Contents

VII Foreword
Mecheels, S. (Boennigheim)

IX Preface
Elsner, P., Hipler U.-C. (Jena)

Interactions between Skin and Textiles

1 Skin Physiology and Textiles – Consideration of Basic Interactions
Wollina, U. (Dresden); Abdel-Naser, M.B. (Cairo); Verma, S. (Baroda)

Interactions between Skin and Biofunctional Metals

17 Silver in Health Care: Antimicrobial Effects and Safety in Use
Lansdown, A.B.G. (London)

Efficiency of Biofunctional Textiles

35 Antimicrobials and the Skin Physiological and Pathological Flora
Elsner, P. (Jena)

42 Antimicrobial Textiles – Evaluation of Their Effectiveness and Safety
Höfer, D. (Boennigheim)

51 Physiological Comfort of Biofunctional Textiles
Bartels, V.T. (Boennigheim)

Safety Evaluation of Biofunctional Textiles

67 Antimicrobial Textiles, Skin-Borne Flora and Odour
Höfer, D. (Boennigheim)

78 Hygienic Relevance and Risk Assessment of Antimicrobial-Impregnated Textiles
Kramer, A. (Greifswald); Guggenbichler, P. (Erlangen); Heldt, P.; Jünger, M.; Ladwig, A.; Thierbach, H.; Weber, U.; Daeschlein, G. (Greifswald)

Manufacturing of Biofunctional Textiles

110 Production Process of a New Cellulosic Fiber with Antimicrobial Properties
Zikeli, S. (Frankfurt/Main)

Biofunctional Textiles in the Prevention and Treatment of Skin Diseases

127 Use of Textiles in Atopic Dermatitis.
Care of Atopic Dermatitis
Ricci, G.; Patrizi, A.; Bellini, F.; Medri, M. (Bologna)

144 Coated Textiles in the Treatment of Atopic Dermatitis
Haug, S.; Roll, A.; Schmid-Grendelmeier, P.; Johansen, P.; Wüthrich, B.; Kündig, T.M.; Senti, G. (Zürich)

152 Silver-Coated Textiles in the Therapy of Atopic Eczema
Gauger, A. (München)

165 A New Silver-Loaded Cellulosic Fiber with Antifungal and Antibacterial Properties
Hipler, U.-C.; Elsner, P.; Fluhr, J.W. (Jena)

179 Antimicrobial-Finished Textile Three-Dimensional Structures
Heide, M.; Möhring, U. (Greiz); Hänsel, R.; Stoll, M. (Freiberg); Wollina, U.; Heinig, B. (Dresden-Friedrichstadt)

200 Author Index

201 Subject Index

Foreword

Biofunctional textiles present a novel disciplinary and scientific field. It evolved through the need to create very specific, biologically functional materials which would have a targeted efficacy on human skin.

At first, the experts were very skeptical as to whether it could be possible to change the structure of a textile, especially the surface of the fibers, to a point where it would be able to take over the biological functions of the skin. However, thanks to the research foundations on the mechanisms of surface kinetics and of the forming of depot structures, it took a relatively short time to reach a promising approach for the task. With the founding of the Competence Center for Textiles and Skin in 2002, research data were gathered from the Clinic for Dermatology and Dermatological Allergology at the Clinic of the Friedrich Schiller University at Jena, the German Textile Research Center North-West, Krefeld, and the Hohensteiner Institute at Boennigheim, and a new scientific field was introduced, namely 'biofunctional textiles'.

Experts in textiles and medicine acknowledged which new possibilities lay in producing biofunctional textiles, especially their functional properties. Research in dermatology and clinical practice were already at a very advanced stage and the time had come for the vast knowledge accumulated by the individual researchers and research groups to be brought together. This has been achieved in this book. Thanks go to the initiators, Prof. Dr. med. Peter Elsner and Dr. rer. nat. Uta-Christina Hipler, for making this state-of-the-art expertise available to the public. Leading experts in the fields of textiles and medicine are highly appreciative of this achievement.

The current state of knowledge forms a good basis for research into functionally useful products. We hope that the scientific knowledge published herein will encourage a more objective discussion on biofunctional textiles and the weighing up of emotional objections against scientific argumentation.

With biofunctional textiles, the textile industry and medicine have taken a step forward together along the path to enriching the use of textile materials for the good of mankind.

Stefan Mecheels, Boennigheim

Preface

According to the archaeologists and anthropologists, the earliest clothing probably consisted of fur, leather, leaves or grass, draped, wrapped or tied about the body for protection from the elements. Knowledge of such clothing remains inferential, since clothing materials deteriorate quickly compared to stone, bone, shell and metal artifacts. Anthropologists at the Max Planck Institute for Evolutionary Anthropology have conducted a genetic analysis of human body lice that indicates that they originated not more than about 72,000 ± 42,000 years ago. Since most humans have very sparse body hair, body lice require clothing to survive, so this suggests a surprisingly recent date for the invention of clothing. Its invention may have coincided with the spread of modern *Homo sapiens* from the warm climate of Africa, thought to have begun between 50,000 and 100,000 years ago.

The significance of clothing is extensive, including clothing as a social message. Social messages sent by clothing can involve e.g. social status, occupation, ethnic and religious affiliation, marital status or sexual availability. Anyway, the practical functionality of clothing is the most important feature.

Practical functions of clothing include providing the human body protection against the weather – strong sunlight, extreme heat or cold, and rain or snow – also protection against insects, noxious chemicals, weapons and contact with abrasive substances. In sum, clothing protects against anything that might injure the naked human body. Humans have shown extreme inventiveness in devising clothing solutions to practical problems.

Especially in recent years, new technologies have been permitting the production of 'functional textiles' and 'smart textiles', i.e. textiles capable of sensing changes in environmental conditions or body functions and responding to these changes. The examples of special fabrics cover underwear with integrated cardio-online system up to textiles with carrier molecules. Such fabrics are able to absorb substances from the skin or can release therapeutic or cosmetic compounds to the skin.

The current interest in biofunctional textiles is mainly focussed on the use of such textiles supporting therapy and prevention in dermatology.

Textiles interact with the skin in a very intensive manner. Therefore, the microorganisms of the skin can influence the skin itself, the textiles as well as the interaction between skin and textiles. During the last few years, the materials for manufacturing textiles show positive tendencies towards a higher functionality. The market has been enriched with innovative antimicrobial products, especially with silver fibers or materials with enclosed silver ions. These textiles could not only find a domain in the wellness sector, but the goal is to use textile fabrics with antimicrobial finishing sufficient for prophylaxis and therapy.

On the other hand wearing these new textiles can generate problems, unknown till now. Potential health risks can occur. To minimize such risks, careful and reliable in vitro as well as in vivo test systems should be established, which is, by the way, one of the most important requirements of the European Conference on Textiles and Skin. Standards are necessary for the effectiveness of antimicrobial textiles as well as for the evaluation of their undesirable side effects, like cytotoxicity, allergenic and irritative potentials.

Because of the fact that this subject is of current interest, many papers have been published in the last few years about the interaction between textiles and skin. Also, a previous volume in this series, *Textiles and Skin*, was well accepted by researchers, dermatologists and others interested in learning about this important subject.

Therefore, the editors decided to continue this successful project. This volume in the series *Current Problems in Dermatology* collects information about the new trends in the interaction of textiles and skin and especially the development of antimicrobial-finished textiles. We apologize that not all aspects of this topic could be taken into consideration and trust that all readers will accept the choice we made. Hopefully, this issue will contribute to the further consolidation of the dialogue between dermatologists and textile engineers.

The editors thank all the authors for their effort contributing to this volume with articles of excellent quality. Finally, we would like to thank the staff of S. Karger AG, Basel, for the productive co-operation and their kind help with this project.

Uta-Christina Hipler
Peter Elsner
Jena, 2006

Burg G (ed): Biofunctional Textiles and the Skin.
Curr Probl Dermatol. Basel, Karger, 2006, vol 33, pp 1–16

Skin Physiology and Textiles – Consideration of Basic Interactions

U. Wollina[a], *M.B. Abdel-Naser*[b], *S. Verma*[c]

[a]Department of Dermatology, Hospital Dresden-Friedrichstadt, Academic Teaching Hospital of the Technical University of Dresden, Dresden, Germany; [b]Department of Dermatology, Ain Shams University Cairo, Cairo, Egypt; [c]Private Practice, Baroda, India

Abstract

The skin exerts a number of essential protective functions ensuring homeostasis of the whole body. In the present review barrier function of the skin, thermoregulation, antimicrobial defence and the skin-associated immune system are discussed. Barrier function is provided by the dynamic stratum corneum structure composed of lipids and corneocytes. The stratum corneum is a conditio sine qua non for terrestrial life. Impairment of barrier function can be due to injury and inflammatory skin diseases. Textiles, in particular clothing, interact with skin functions in a dynamic pattern. Mechanical properties like roughness of fabric surface are responsible for non-specific skin reactions like wool intolerance or keratosis follicularis. Thermoregulation, which is mediated by local blood flow and evaporation of sweat, is an important subject for textile-skin interactions. There are age-, gender- and activity-related differences in thermoregulation of skin that should be considered for the development of specifically designed fabrics. The skin is an important immune organ with non-specific and specific activities. Antimicrobial textiles may interfere with non-specific defence mechanisms like antimicrobial peptides of skin or the resident microflora. The use of antibacterial compounds like silver, copper or triclosan is a matter of debate despite their use for a very long period. Macromolecules with antimicrobial activity like chitosan that can be incorporated into textiles or inert material like carbon fibres or activated charcoal seem to be promising agents. Interaction of textiles with the specific immune system of skin is a rare event but may lead to allergic contact dermatitis. Electronic textiles and other smart textiles offer new areas of usage in health care and risk management but bear their own risks for allergies.

The skin is a communicative, sensitive and protective organ. The skin surface with the stratum corneum represents a critical structure in the interaction of the human body with the environment. Without the horny layer a terrestrial life would be impossible. On the other hand, the stratum corneum becomes impaired in any kind of superficial or deep injuries. To retain body homeostasis, recovery of the stratum corneum is necessary.

Other important protective functions of the skin are protection against infection and irradiation, in particular ultraviolet irradiation, thermoregulation and synthesis of hormones and other bioactive substances. The skin has a great importance in social life and is a basis of attractiveness as well. Many of these functions are also related to an intact barrier function.

The epidermis derives from the ectoderm. Initially the epidermis comes as a monolayer. In the first trimester the epidermis is covered by a single-layered periderm. Epidermal stratification starts after 8 weeks. During the second trimester cornification is realized. Then the periderm disappears and becomes a constituent of the vernix caseosa. The epidermal thickness in an immature newborn is about 29 μm, in mature newborns and adults the thickness is about 50 μm. The skin surface of the newborn is covered by the protective gelatinous vernix caseosa, whereas the skin surface of adults is rather dry [1].

If one considers clothing as a protective measure of skin in human life, the vernix caseosa may be seen as the first 'clothing' in an individual's life produced by the mother's body to protect the newborn especially during the very first period of adaptation to terrestrial life.

Stratum Corneum Barrier and Skin Surface

The stratum corneum is the essential structure of the skin barrier. The major constituents of the stratum corneum are lipids and proteins. In a typical case about 20 layers of nucleus-free corneocytes densely packed with keratin filaments are surrounded by a matrix. The matrix is composed of filaggrin and its derivates, and lipid-rich lamellar bodies. Lamellar bodies fuse end to end, thereby forming lipid double layers [2].

The lipid constitutes are cholesterol, ceramides and free fatty acids. Ceramides stand for about 50% of horny layer lipids and are essential for the lamellar structure of the epidermis. Cholesterol regulates the phase behaviour of the stratum corneum. The free fatty acids are mostly long-chained molecules with more than 20 C atoms. Lipids are responsible for the hydrophobicity of the horny layer [3]. The water exchange occurs via migrating pores, i.e. polar transport pathways within the lipid mosaic [4]. The matrix develops under the influence of pH gradients, sodium ions and enzymes (synthetases, reductases, hydrolases, lipases) [3]. In a simplistic way, the stratum corneum can be described by the bricks-in-mortar model [2]. More recent studies discovered subunits of octahedrons of corneocytes within the horny layer which may migrate [4].

The horny layer is covered on its surface by a thin amorphous film contributing to stratum corneum structure and function. In newborns there is an almost neutral skin surface with a pH of 6.6 that changes within days or weeks into an acidic pH of 5.9 (acid skin surface film). This leads to activation of pH-dependent hydrolytic enzymes like β-glucocerebrosidase and stratum corneum secretory phospholipase A_2 [5, 6]. Skin surface pH is modulated by microbial harvest, eccrine and sebaceous gland secretions, and endogenous catabolic pathways. The acidification of the horny layer is necessary for barrier function [7]. Exposure of the horny layer to neutral buffers (i.e. wet work conditions) or blocking of acidification increases the pH. Hereby, serine proteases become activated that digest desmoglein 1. Desmoglein 1 is a major constituent of corneosomes. Metabolization of desmoglein 1 decreases cohesivity of corneocytes and enforces horny layer permeability [8].

The horny layer impairment as subclinical dryness of skin is quite common. It may have substantial impact on the whole body. Itching sensations are a common symptom [9]. Prolonged exposure of human skin to wetness (water) and/or occlusion leads to measurable disturbances of barrier function. The transepidermal water loss increases in relation to duration of exposure and temperature [10, 11].

Interaction of clothing with the skin surface is first a mechanical one with the skin surface structure. Friction and pressure are the major forces. Surface quality of textiles may directly interfere with skin integrity. This is of particular importance for socks. The plantar skin is not only the thickest of the whole human body with a well-developed multilayered stratum corneum, it has to stand repeated pressure and friction during the whole human life. Even the smallest peaks of pressure when occurring frequently enough – as caused by a seam for instance – may cause skin lesions in patients with impaired skin resistance (e.g. bullous disease or diabetes) [12]. Friction is also a major cause of the induction of keratosis follicularis on the thighs and the outer upper arms [13].

Xerosis cutis is a consequence of the reduction of epidermal water content (<10% of stratum corneum). An increased transepidermal water loss leads to itching, scaling, roughness and fissuring [14]. Xerosis might be a symptom of irritant contact dermatitis, atopic dermatitis or ichthyosis. In case of so-called sensitive skin which is often a dry skin as well, e.g. in atopic dermatitis, the tactile threshold is lowered. Rough textile surfaces, such as wool, can induce irresistible prickling and itching known as 'wool intolerance' [13, 15]. Smooth textile surfaces are often more comfortable irrespective of the fibres included. These qualities are of particular relevance in direct skin-textile interaction like in underwear. The perceived importance of fabrics and sweat as triggering exacerbating factors in atopic dermatitis is high. About 40% of 12- to 14-year-old schoolchildren believe that wool fabrics and sweating during exercise are worsening their skin condition [16].

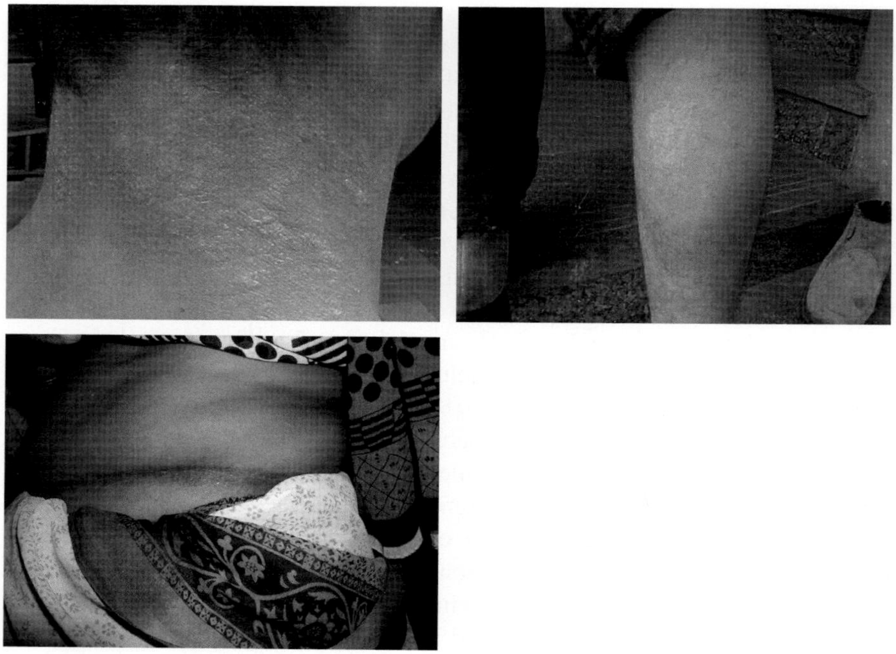

a b c

Fig. 1. Mechanical irritation. ***a*** Mechanical irritation by clothing of the polyamide type; cotton was well tolerated. The lichenoid lesions healed only by changing the clothing. ***b*** Hair loss induced by tight polyester jeans. ***c*** The traditional sari friction dermatosis causes hyperpigmentation and lichenoid papules in the hip region.

Atopic dermatitis and dry skin often flare up during wintertime, but the mechanisms of winter deterioration of dry and atopic skin are not fully understood. Dryness of textile-protected skin in particular is prominent around the shoulders (fig. 1). Change of washing clothing with anionic, additive-enriched detergents to a non-ionic, additive-reduced detergent for a period of 2 weeks improved skin conditions in Japanese patients who had worn cotton underwear and used to wash it with cold tap water. Under these conditions residues of common washing detergents/surfactants in cotton underclothes may contribute to winter worsening of dry or atopic skin [17]. In an experimental study on professional laundry, a broad variety of surfactant residues according to the type and amount could be identified on cotton-based textiles. The skin reaction during patch testing in healthy and skin-sensitive adults was not strictly correlated with a threshold limit or the concentration of surfactants. Non-ionic surfactants were better tolerated than anionic ones [18].

Clothing and Thermoregulation

In neonates body temperature rapidly drops soon after birth. In order to survive, neonates must accelerate heat production by lipolysis in brown adipose tissue, i.e. non-shivering thermogenesis. This process is oxygen dependent [19]. Clothing has to function as insulation and therefore supports thermoregulation. Heat loss prevention is a major task in the delivery room and thereafter, which is of particular relevance in preterm neonates.

The cutaneous thermosensitivity is not evenly distributed over the body. Local thermosensitivity can be calculated from changes in sweat rates and thermal discomfort. The highest cold sensitivity is found on facial skin, 2–5 times higher than in any other part of the body. Facial skin has also the highest warm sensitivity. In contrast, the limb extremities are the least thermosensitive segment for warming and cooling [20].

A key function of clothing is insulation. Thickness of the material and therefore the volume of air enclosed in the fabric appears to be the major determinant. Dry heat transfer through fabrics consists mainly of conduction and radiation. During exercise or in extremes of environment (cold and heat), the interaction of body thermoregulation with clothing gains even greater importance. Recent developments with phase-changing materials open the opportunity for buffering heat [21].

The total body as well as per body surface sweating rate increases after puberty. In contrast mass-related evaporative cooling and sweating efficiency are highest in prepubertal humans [22]. These findings illustrate the different needs of age groups in terms of thermoregulative support by clothing. Although sweat gland activity is directly controlled by the central nervous system controlling the core body temperature, sweat glands can also be influenced by local cutaneous thermal conditions. It might be considered that improper clothing can support the phenotypic realization of focal (axillary or plantar) hyperhidrosis. Local temperatures above 32°C predominantly affect neurotransmitter release [23]. On the other hand, there are also gender- and gene-related differences in thermoregulation of humans [24]. The sweating rate is higher in males [25]. In women after the menopause, changes in reproductive hormone levels substantially alter the thermoregulatory control of skin blood flow as illustrated by the occurrence of hot flashes. Pre-existent pathological conditions alter the thermoregulative response as well. In type 2 diabetics the ability of skin blood vessels to dilate is impaired [26]. Compared to normal males hypertensive males develop higher skin temperatures in the heat. Water ingestions recommended for normal men during exercise may cause abnormal cardiac workload in hypertensive individuals [27]. Functional textiles may support patients to ensure a quite normal body temperature without cardiac overload by water ingestion.

Cold stress can quickly overwhelm human thermoregulation leading to impaired performance or even death. Convective heat loss is the most important factor. Cold exposure induces both vasoconstriction and thermogenesis [28]. Several layers of textiles with thermoinsulative qualities and a wet-protective outer surface are capable of allowing to stay in the cold and an unfriendly environment.

Clothing at its best is an interactive barrier ensuring thermal balance despite changes in ambient temperature and humidity, metabolic heat production, gender and age differences, and intended use [29, 30]. In such a way clothing can serve a protective function by reducing radiant heat gain and thermal stress. Clothing construction affects these functions [31].

During exercise in moderate heat, a clothing fabric that promotes sweat evaporation did not affect mean body temperature, rectal temperature (as a measure of core temperature), mean skin temperature, heart rate or comfort sensation responses compared to a traditional cotton fabric [32]. The more intense the exercise and/or the more extreme the environment, the higher is the impact of clothing on thermoregulation [33]. Intermittent regional microclimate cooling is more than twice as efficient in reducing exercise heat strain than constant microclimate cooling [34]. Adoptive functions of textiles should be capable of supporting microclimate changes. Major items of interaction of clothing with the human body and the environment are thermal and water vapour resistance, mass transfer, directed fluid transport (e.g. of sweat), evaporation, thermal load, air gaps and contact layers. Perception of clothing comfort is positively related to warmth and negatively to dampness [35].

Clothing and the Skin-Associated Antimicrobial Defence System

The skin-associated immune system comprises specific and non-specific defence mechanisms. The non-specific ones will be discussed in detail, since they seem to represent the base for human body homeostasis. The specific part is composed of antigen-presenting cells (i.e. Langerhans cells and epidermal keratinocytes) and lymphocytes (B, T and NK types).

The prototype of a specific immune reaction to clothing, completely unwanted but fortunately quite rare, is (allergic) contact dermatitis. The induction of an allergic contact dermatitis has been recognized since the 19th century. This type of an allergic contact dermatitis needs an intense contact with the skin surface. Areas with a high sweat gland density are at special risk such as the intertriginous skin. The most common allergens are textile dyes such as disperse dyes [36, 37]. Rarely other components like fibre additives, finishing or contaminants may be responsible for textile-related allergies. For a more

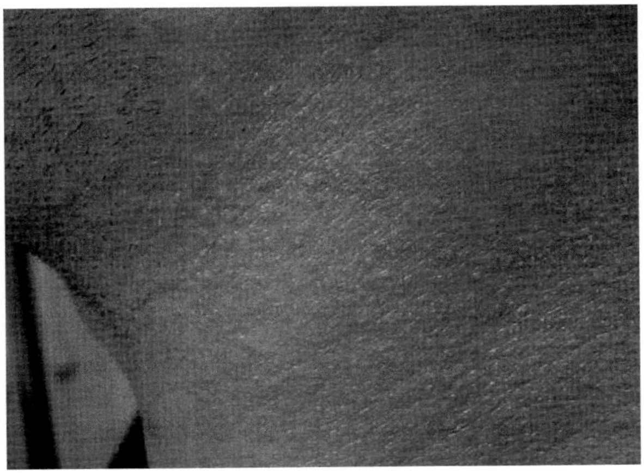

Fig. 2. Acute contact dermatitis due to clothing and sweating. Sweat is a critical component in interactions between human skin and textiles. Sweat may dissolve dyes and other components from clothing and increase the skin permeability for hydrophilic compounds.

detailed discussion we would like to refer to the excellent review by Hatch and Maibach [36] and Le Coz [37] (fig. 2). The matter needs a re-evaluation with development and usage of smart, hybrid and interactive or electronic textiles in the market [38–40]. However, even more important is the possible interference of textiles with non-specific defence mechanisms of the skin.

The non-specific immune function of skin involved in antimicrobial defence is of even greater importance, provides the base for resistance to microbial threat and may interact in several ways with the specific immune function whenever necessary.

The stratum corneum is the essential structure for non-specific resistance supported by skin gland secretions. There is a close relationship between horny layer barrier function and the risk of skin infection [41]. In recent years research has identified several families of antimicrobial peptides in vertebrates including humans [42, 43]. Many of these peptides are multifunctional. They are not only natural antibiotics but chemotaxins as well. The human cathelicidin LL-37 is chemotactic for neutrophils, monocytes, mast cells and T lymphocytes, causes mast cell degranulation and supports vascularization and re-epithelialization of wounds [44].

In human skin appendages genes like *DCD* and *CAMP* are expressed encoding antimicrobial peptides cathelicidin LL-37 and dermicidin. By the action of serine proteases, new antimicrobials are produced with their own antimicrobial profile [45, 46]. β_2-Defensin has been identified in lamellar bodies of the human skin [47]. There seems to be some antimicrobial activity of the lipid phase in the stratum corneum. In addition, such substances are secreted together with the sweat and spread on the skin surface [46, 48].

In newborns there is a 10- to 100-fold increase in the expression of cathelicidin LL-37 and β_2-defensin compared with adult skin [49]. This can be viewed as a compensatory mechanism of a still immature immune system in adapting to postuterine life. The antimicrobial peptides are concentrated in the vernix caseosa. Both α_1- to α_3-defensin and cathelicidin LL-37 have been detected there [50].

During the recovery of barrier function after wounding or inflammatory disease (eczema, psoriasis) there is a close interaction between growth factors and antimicrobial peptides [51]. The microbial settlement induces antimicrobial control mechanisms: for instance, the saprophytic yeast *Malassezia furfur*, but bacteria as well, induces the expression of β_2-defensin in human epidermal keratinocytes [52, 53].

Clothing may impair the antimicrobial defence of skin by mechanical alteration of the barrier function but also by its effects on skin wetness and local blood flow. Athletes are prone to develop tinea pedis, 'athletes' feet', due to the humidity in socks and sports shoes. Viral plantar warts are not uncommon and their spread can be facilitated by plantar hyperhidrosis [54].

On the other hand, textiles can support the non-specific cutaneous defence. In the most simplistic way, clothing provides a mechanical barrier against infestation, insect bites, protozoa and microbes, e.g. the mosquito net or the protective textiles in an operation room.

The use of textile materials to support and deliver active chemicals is as old as the application of an ointment to a fabric for covering a wound. What may not be so obvious is the potential volume and number of textile products which could benefit from this technology. Medical products are perhaps the largest application of this kind of technology. In health-related professions, protection from pathogens is a growing concern, and textiles with antimicrobial properties are desirable. Fungi, bacteria and associated insects are responsible for significant infections and allergy problems. Less obvious applications for antimicrobial textiles include air filters (indoor air quality), carpets, draperies, wall coverings etc., particularly in environments where the sick, elderly and other susceptible individuals live.

Until recently, biocidal textiles have not been widely available in the market, in spite of the obvious commercial potential. Even with the recent announcements of new products, their efficacy and safety are not readily compared. There is a reasonable array of chemicals which can act as biocides, including chemical oxidants, photo-oxidants, membrane disrupters, heavy metals, organic protein denaturants and chemicals which mimic biochemical intermediates but which function improperly in the micro-organism. Protein denaturants and biochemical metabolites are likely to require release from the biocidal fibre and diffusion through the cell wall before being effective. Membrane disrupters may work acceptably from outside the microbe and may not require release from the fibre, at least until the microbe is in close proximity [55].

General oxidants and some membrane disrupters appear to be the least objectionable and offer the greatest potential as biocidal agents in textiles. The more specific materials which inhibit micro-organisms by malfunctioning in a metabolic pathway are more elegant, but are more likely to induce immunity in the organism. Induced immunity is a problem which has just been reported for the latest, widely used antimicrobial agent for consumer goods.

The scientific and technological approaches are complex. Biocidal particles have to be attached to the fibre surface or alternatively incorporated into the fibre. Conventional chemical reactions are evaluated to attach molecules and particles covalently to the fibre surface, but also copolymerization of biocidal monomers into the fibre-forming polymer – followed by fibre extrusion, surface grafting of a biocidal monomer covalently to the fibre surface with ultraviolet or peroxide initiation, and incorporation of an antimicrobial agent into fibre during extrusion.

Finally, the molecular mechanics modelling approaches are under investigation. They have to address each of the issues such as the number of reactive groups which can be stored on the surface, the kinetics of ionic release and the mobility of the surface-grafted molecules.

Although it would seem that the number of reactive species that could be stored by simply grafting a 'storage polymer' to the surface of the fibre could increase without bound, available data argue against this. When the radius of gyration of the storage polymer is equal to one half the average distance between graft sites, the maximum storage potential is achieved. For smaller molecules, the surface has many bare spots which do not store any reactive species. When the storage polymer is larger than optimum, it blocks adjacent graft sites, thus preventing attachment of additional storage polymers, again resulting in bare fibre surface. Recent investigations try to optimize the effect of polydispersity, both in the storage polymer and in the graft site density on the surface, on storage efficiency.

Most polymers (except cellulose) have very few reactive sites on their surfaces. Therefore, it is impractical to store a large amount of reactive species on their surfaces unless an amplification system is used. Although plasma treating can be used, it is too expensive for most applications. Another approach is to graft to the surface a polymeric material that contains groups to which the reactive species can be grafted, e.g. poly(acrylic acid), poly(vinyl alcohol), poly(vinyl amine) or copolymers containing these groups [55]. There are several issues that need to be dealt with: (a) What is the maximum number of reactive groups that can be stored on the surface? (b) How can they be released? (c) How does the mobility of the surface-grafted polymer affect the delivery of the reactive species?

The durability or rechargeability of the fabric is a practical problem. Recharging would reduce waste products. Can it be realized in situ or does it need repeated processing of the fabric? More important for textile-skin interaction are fundamentals of safety and toxicity. Under these circumstances the release of active chemicals from the fabric bears a greater potential risk than biocidals that remain attached either by covalent binding or physical attachment to the textile. Novel approaches include inclusion of biocidal compounds into silica matrices using the sol-gel technique. Moreover, silica coatings with embedded nanoparticular silver combined with organic biocidal compounds effectively decreased the survival rates of different bacteria on textiles and medical catheters [56].

To enhance the protective efficiency against insects, clothing has been impregnated with repellents. As shown in a French study in Côte d'Ivoire on military health service personnel, permethrin impregnation of uniforms improved protection against mosquito bites but not enough to reduce significantly the incidence of malaria among non-immune troops [57].

Copper and silver ions exert antibacterial, fungicide and nematocide activities. They are not skin sensitizing. Their use in mattresses, antifungal socks, anti-dust-mite mattress covers or antibacterial fabrics is under investigation [58, 59]. Silver-nylon fibres were effective in vitro against *Staphylococcus aureus*, *Pseudomonas aeruginosa* and *Candida albicans*. In vivo silver-nylon cloth prevented colonization of burn wounds [58]. Silver-coated textiles have been shown to reduce colonization of skin by *S. aureus* in atopic dermatitis within a couple of days [60]. The *S. aureus* cell wall compononents peptidoglycan and lipoteichoic acids have been identified to activate human keratinocytes by activation of toll-like receptor 2 pathways [61]. In the clinical study reduction of skin colonization was accompanied by clinical improvement of atopic dermatitis [60].

A combination of polyhexamethylene biguanidine, quaternary ammonium silane organic-based compounds and silver was developed for uniform bacteria mitigation on the military battle field. After 24 h a 100% effectiveness was obtained against Gram-positive bacteria on nylon-based fabrics [62]. Heavy metals and protein denaturants are effective, but toxicity problems are a concern [63].

Silver is used as a topical agent in wound dressings for burns, diabetic ulcers and other chronic open wounds. Absorption can lead to deposition within the wound and internal organs such as the liver and kidney. Despite this, the risk of lasting tissue damage or functional disorders has been estimated to be low [64].

Very recently, de novo synthesis of biomimetic polymers and oligomers that mimic the structure and biological activity of natural antimicrobial peptides such as margainins and creopins has become reality. It remains to be seen whether such compounds will offer new options and safety in finished textiles.

The resident skin flora represents a balanced ecosystem, where resident germs play an important physiological role. One of the main benefits that humans derive from resident flora is protection from infection. Any change in the microbiological equilibrium can carry negative consequences [65]. In preterm infants, coagulase-negative staphylococci are the most common species on the skin representing about 80% of the neonate's flora, numerous strains with antibiotic resistance or multiresistance [66]. Later on there is a great interindividual and intra-individual variation of resident microflora. With increasing age, streptococci disappear and corynebacteria occur. Anaerobic propionibacteria are more frequent in youngsters when sebum production is increased. Micrococci, Gram-negative germs like *Acinetobacter* and yeasts like *Malassezia* spp. are also components of the resident flora [67].

The antibacterial activity of biocide-finished textile products containing silver, zinc, ammonium zeolite and chitosan was evaluated under wet and dry conditions. The antibacterial activity was limited to wet conditions. Addition of organic matter decreased the antibacterial activity. In any case antibacterial effects required several hours of incubation. Some bacteria species and strains were not affected. The authors concluded that antibacterial properties of biocide-finished textiles in the clinical setting may be of limited value [68].

The body odour is a product of sweat gland activity, bacterial contamination (in particular corynebacteria) and steroidal metabolism. Diseases (like diabetes, cancer or Fish odour disease) interfere with body odour just as nutrition or medication and the use of cosmetics and hygiene products as well. Body odour may be a signal of sexual attraction and a non-verbal communicator of emotions [69, 70]. Malodour of the axillary and pubic region has been associated with *Corynebacterium* spp. which are capable of generating several odorous compounds derived from androgens [71]. Reduction of malodour can be achieved by antimicrobial measures and reduction of sweat gland activity. Textiles with a high rate of directed vapour transport may thereby reduce the odour in a non-specific way. Antimicrobial activity is used to diminish the bacterial load. The antibacterials should be fixed to the textile fibres to avoid contact sensitization and disturbances of non-specific microbial defence by the resident microflora. Development of bacterial resistance has to be considered as a serious problem [72, 73].

Prevention of odour and discoloration of the textile material are significant, if less critical reasons to use antimicrobials. Consumers are showing increasing interest in antimicrobial products, particularly those products like carpets which are suspected of harbouring microbes. The applications will result in products which will become pervasive throughout the medical services, filtration industries, home furnishings and selected apparel items (like socks). Indeed, several product introductions in this area have been made.

Triclosan-incorporated polymers are on the market for hospital use as fabric seat covers, chairs and clothing. Testing antibacterial activity on the other hand was found to be discouraging. In light of recent studies that have shown specific interactions of triclosan with the bacterial lipid synthesis pathway, triclosan-incorporated polymers may provide an ideal setting for resistant strains of bacteria to grow and thus should not be used in a broad range but only in selected hospital settings [74]. Safety concerns have been addressed on the other hand because of a possible interaction with the normal skin flora, percutaneous absorption and systemic toxicity [73]. Human studies suggest that percutaneous absorption of heavy metals through intact skin is poor [64].

The situation is somewhat different for triclosan. After in vivo topical application of a 64.5-m*M* alcoholic solution of [^{3}H]triclosan to rat skin, 12% radioactivity was recovered in the faeces, 8% in the carcass, 1% in the urine, 30% in the stratum corneum and 26% was rinsed from the skin surface 24 h after application. Free triclosan and the glucuronide and sulfate conjugates of triclosan were found in urine and faeces. Triclosan penetrated rat skin more rapidly and extensively than human skin in vitro. Twenty-three percent of the dose had penetrated completely through rat skin into the receptor fluid by 24 h, whereas penetration through human skin was only 6.3% of the dose. Chromatographic analysis of the receptor solutions showed that triclosan was metabolized to the glucuronide, and to a lesser extent to the sulphate, during passage through the skin. Triclosan glucuronide appeared rapidly in the receptor fluid whereas triclosan sulphate remained in the skin. Although the major site of metabolism was the liver, conjugation of triclosan in skin was also demonstrated in vitro and in vivo, particularly to the glucuronide conjugate which was more readily removed from the skin. By extrapolation of the comparative in vitro data for human and rat skin it is reasonable to deduce that dermal absorption in humans of triclosan applied at the same dose is about one third of that in the rat in vivo [75].

Charcoal-containing textiles are in use as wound dressings for the malodorous wound. Creating an enlargement of the adsorptive surface by charcoal, physical binding of debris and bacteria is supported. The health hazards on intact skin are not known. In open wounds carbon may be deposited within the phagocytic cells without further medical problems [64].

Charcoal-containing devices can also be used to improve body odour or to reduce flatus odour when worn inside underwear. Here they are quite efficient to bind sulphide gases when used in construction of briefs made of activated charcoal fibres. Pads were much less effective under this view [76].

Some naturally occurring polymers like chitosan, alginates or kapok fibres offer antimicrobial activity and biocompatibility [77]. Supramolecular structures fixed to textile fibres like cyclodextrins allow to reduce bacterial contamination of sweat-gland-rich body parts like the axillary region or the feet [77]. The importance of laundering in the prevention of skin infections has been discussed in detail elsewhere [78].

Conclusions

Antimicrobial-finished textiles have just entered the market. Antimicrobial-finished textiles for special purposes such as the operation room or the military battlefield seem to offer advantages. Many of the currently available products interfere more intensely with skin-derived microbes on the textile than with the skin-associated microbial ecosystem. The use of soluble or volatile antimicrobials in clothing can be associated with problems such as intolerance of the human body, allergies or disturbances of the human skin resident flora. The development of new technologies and products should seriously consider the delicate balance of the skin microflora and take it into account for the selection of compounds and techniques. Other important areas of interference of textiles with human skin physiology are barrier function of skin and thermoregulation. If the skin is seen as a complex and adaptive organ, textiles and clothing can be created that will support body function and allow keeping homeostasis even in the most unfriendly environment.

References

1 Loomis CA, Birge MB: Foetal skin development; in Eichenfield LF, Frieden IJ, Esterly NB (eds): Textbook of Neonatal Dermatology. Philadelphia, Saunders, 2001, pp 1–17.

2 Elias PM: Epidermal lipids, barrier function, and desquamation. J Invest Dermatol 1993;80 (suppl):44S–49S.

3 Feingold KR: The regulation and role of epidermal lipid synthesis. Adv Lipid Res 1991;24:57–82.

4 Forslind B: A domain mosaic model of the skin barrier. Acta Derm Venereol 1999;74:72–77.

5 Behne MJ, Barry NP, Hanson KM, Aronchik I, Clegg RW, Gratton E, Feingold K, Molleran WM, Elias PM, Mauro TM: Neonatal development of the stratum corneum pH gradient: localization and mechanisms leading to emergence of optimal barrier function. J Invest Dermatol 2003;120: 998–1006.

6 Fluhr JW, Behne MJ, Brown BE, Moskowitz DG, Selden C, Mao-Qiang M, Mauro TM, Elias PM, Feingold KR: Stratum corneum acidification in neonatal skin: secretory phospholipase A_2 and the sodium/hydrogen antiporter-1 acidify neonatal rat stratum corneum. J Invest Dermatol 2004;122: 320–329.

7 Chapman SJ, Walsh A: Membrane-coating granules are acidic organelles which possess proton pumps. J Invest Dermatol 1989;93:466–470.

8 Hachem JP, Crumrine D, Fluhr J, Brown BE, Feingold KR, Elias PM: pH directly regulates epidermal barrier homeostasis, and stratum corneum integrity/cohesion. J Invest Dermatol 2003;121: 345–353.

9 Yoipovitch G: Dry skin and impairment of barrier function associated with itch – New insights. Int J Cosmet Sci 2004;26:1–7.
10 Hildebrandt D, Ziegler K, Wollina U: Electrical impedance and transepidermal water loss of healthy human skin under different conditions. Skin Res Technol 1998;4:130–134.
11 Kligman AM: Hydration injury to human skin: a view from the horny layer; in Kanerva L, Elsner P, Wahlberg JE, Maibach HI (eds): Handbook of Occupational Contact Dermatitis. Berlin, Springer, 2000, pp 76–80.
12 Wollina U: Der diabetische Fuss – Eine Übersicht für Dermatologen. Z Hautkrankh 1999;74: 265–270.
13 Tronnier H: Wirkungen von Textilien an der menschlichen Haut. Dermatol Beruf Umwelt/Occup Environ Dermatol 2002;50:5–10.
14 Grubauer G, Elias PM, Feingold KR: Trans-epidermal water loss: the signal for recovery of barrier structure and function. J Lipid Res 1989;30:323–333.
15 Fisher AA: Non-allergic 'itch' and 'prickly' sensation to wool fibres in atopic and non-atopic persons. Cutis 1996;58:323–324.
16 Williams JR, Burr ML, Williams HC: Factors influencing atopic dermatitis – A questionnaire survey of schoolchildren's perceptions. Br J Dermatol 2004;150:1154–1161.
17 Kiriyama T, Sugiura H, Uehara M: Residual washing detergent in cotton clothes: a factor of winter deterioration of dry skin in atopic dermatitis. J Dermatol (Tokyo) 2003;30:708–712.
18 Matthies W: Tensidrückstände auf gewerblicher Wäsche und ihre Bedeutung für die dermatologische Verträglichkeitsbewertung. Dermatol Beruf Umwelt/Occup Environ Dermatol 2001;49: 102–107.
19 Asakura H: Foetal and neonatal thermoregulation. J Nippon Med Sch 2004;71:360–370.
20 Cotter JD, Taylor NA: The distribution of cutaneous sudomotor and alliesthesial thermosensitivity in mildly heat-stressed humans: an open-loop approach. J Physiol 2005;565:335–345.
21 Havenith G: Clothing and thermoregulation; in Elsner P, Hatch K, Wigger-Alberti W (eds): Textiles and the Skin. Curr Probl Dermatol. Basel, Karger, 2003, vol 31, pp 35–49.
22 Inbar O, Morris N, Epstein Y, Gass G: Comparison of thermoregulatory responses to exercise in dry heat among pre-pubertal boys, young adults and older males. Exp Physiol 2004;89: 691–700.
23 De Pasquale DM, Buono MJ, Kolkhorst FW: Effect of skin temperature on the cholinergic sensitivity of the human eccrine sweat gland. Jpn J Physiol 2003;53:427–430.
24 Nguyen MH, Tokura H: Sweating and tympanic temperature during warm water immersion compared between Vietnamese and Japanese living in Hanoi. J Hum Ergol (Tokyo) 2003;32: 9–16.
25 Rosene JM, Whitman SA, Fogarty TD: A comparison of thermoregulation with creatine supplementation between the sexes in a thermo-neutral environment. J Athletic Training 2004;39:50–55.
26 Charkoudian N: Skin blood flow in adult thermoregulation: how it works, when it does not, and why. Mayo Clin Proc 2003;78:603–612.
27 Ribeiro GA, Rodrigues LO, Moreira MC, Silami-Garcia E, Pascoa MR, Camargos FF: Thermoregulation in hypertensive men exercising in the heat with water ingestion. Braz J Med Biol Res 2004;37:409–417.
28 Stocks JM, Taylor NA, Tipton MJ, Greenleaf JE: Human physiological responses to cold exposure. Aviat Space Environ Med 2004;75:444–457.
29 Pascoe DD, Shanley LA, Smith EW: Clothing and exercise. I. Biophysics of heat transfer between the individual, clothing and environment. Sports Med 1994;18:38–54.
30 Pascoe DD, Bellingar TA, McCluskey BS: Clothing and exercise. II. Influence of clothing during exercise/work in environmental extremes. Sports Med 1994;18:94–108.
31 Gavin TP: Clothing and thermoregulation during exercise. Sports Med 2003;33:941–947.
32 Gavin TP, Babington JP, Harms CA, Ardelt ME, Tanner DA, Stager JM: Clothing fabric does not affect thermoregulation during exercise in moderate heat. Med Sci Sports Exerc 2001;33:2124–2130.
33 Barker DW, Kini S, Bernard TE: Thermal characteristics of clothing ensembles for use in heat stress analysis. Am Ind Hyg Assoc J 1999;60:32–37.
34 Cheuvront SN, Kolka MA, Cadarette BS, Montain SJ, Sawka MN: Efficacy of intermittent, regional microclimate cooling. J Appl Physiol 2003;94:1841–1848.

35 Li Y: Perceptions of temperature, moisture and comfort in clothing during environmental transients. Ergonomics 2005;48:234–248.
36 Hatch KL, Maibach HI: Textiles; in Kanerva L, Elsner P, Wahlberg JE, Maibach HI (eds): Handbook of Occupational Dermatology. Berlin, Springer, 2000, pp 622–636.
37 Le Coz C-J: Clothing; in Rycroft RJG, Menné T, Frosch PJ, Lepoittevin J-P (eds): Textbook of Contact Dermatitis, ed 2. Berlin, Springer, 2001, pp 725–749.
38 Da Rocha AM: Development of textile-based high-tech products: the new challenge. Stud Health Technol Inform 2004;108:330–334.
39 Meinander H, Hinkala M: Potential applications of smart clothing solutions in health care and personal care. Stud Health Technol Inform 2004;108:278–285.
40 Schultze C, Burr S: Market research on garment-based 'wearables' and biophysical monitoring and a new monitoring method. Stud Health Technol Inform 2004;108:111–117.
41 Roth RR, James WD: Microbial ecology of the skin. Annu Rev Microbiol 1988;42:441–464.
42 Nicolas P, Vanhoye D, Amiche M: Molecular strategies in biological evolution of antimicrobial peptides. Peptides 2003;24:1669–1680.
43 Izadpanah A, Gallo RL: Antimicrobial peptides. J Am Acad Dermatol 2005;52:381–390.
44 Zanetti M: Cathelicidins, multifunctional peptides of the innate immunity. J Leukoc Biol 2004;75: 39–48.
45 Mangoni ML, Papo N, Mignogna G, Andreu D, Shai Y, Barra D, Simmaco M: Ranacyclins, a new family of short cyclic antimicrobial peptides: biological function, mode of action, and parameters involved in target specificity. Biochemistry 2003;42:14023–14035.
46 Murakami M, Ohtake T, Dorschner RA, Schittek B, Garbe C, Gallo RL: Cathelicidin anti-microbial peptide expression in sweat, an innate defence system for the skin. J Invest Dermatol 2002;119: 1090–1095.
47 Oren A, Ganz T, Liu L, Meerloo T: In human epidermis, beta-defensin 2 is packed in lamellar bodies. Exp Mol Pathol 2003;74:180–182.
48 Schittek B, Hipfel R, Sauer B, Bauer J, Kallenbacher H, Stevanovic S, Schirle M, Schroeder K, Blin N, Meier F, Rassner G, Garbe C: Dermicidin: a novel human antibiotic peptide secreted by sweat glands. Nat Immunol 2001;2:1133–1137.
49 Dorschner RA, Lin KH, Murakami M, Gallo RL: Neonatal skin in mice and humans expresses increased levels of antimicrobial peptides: innate immunity during development of the adaptive response. Pediatr Res 2003;53:566–572.
50 Yoshio H, Tollin M, Gudmundsson GH, Lagercrantz H, Jornvall H, Marchini G, Agerberth B: Antimicrobial polypeptides of human vernix caseosa and amniotic fluid: implications for newborn innate defense. Pediatr Res 2003;53:211–216.
51 Sorensen OE, Cowland JB, Theilgaard-Mönch K, Liu L, Ganz T, Borregaard N: Wound healing and expression of antimicrobial peptides/polypeptides in human keratinocytes, a consequence of common growth factors. J Immunol 2003;170:5583–5589.
52 Donnarumma G. Paoletti I, Biommino E, Orlando M, Tufano MA, Baroni A: *Malassezia furfur* induces the expression of beta-defensin-2 in human keratinocytes in a protein C-dependent manner. Arch Dermatol Res 2004;295:474–481.
53 Chung WO, Dale BA: Innate immune response of oral and foreskin keratinocytes: utilization of different signalling pathways by various bacterial species. Infect Immun 2004;72:352–358.
54 Adams BB: Dermatologic disorders of the athlete. Sports Med 2002;32:309–321.
55 Edwards JV, Vigo Tl: Bioactive Fibres and Polymers. Weimar/Texas, Culinary and Hospitality Industry Publications Service, 2001.
56 Haufe H, Thron A, Fiedler D, Mahltin B, Böttcher H: Biocidal nanosol coatings. Surf Coatings Int B Coating Trans 2005;88:55–60.
57 Deparis X, Frere B, Lamizana M, N'Guessan R, Leroux F, Lefevre P, Finot L, Hougard JM, Carnevale P, Gillet P, Baudon D: Efficacy of permethrin-treated uniforms in combination with DEET topical repellent for protection of French military troops in Côte d'Ivoire. J Med Entomol 2004;41:914–921.
58 Deitch EA, Marino AA, Malakanok V, Albright JA: Silver nylon cloth: in vitro and in vivo evaluation of antimicrobial activity. J Trauma 1987;27:301–304.

59 Borkow G, Gabbay J: Putting copper into action: copper-impregnated products with potent biocidal activities. FASEB J 2004;18:1728–1730.
60 Gauger A, Mempel M, Schekatz A, Schäfer T, Ring J, Abeck D: Silver-coated textiles reduce *Staphylococcus aureus* colonization in patients with atopic eczema. Dermatology 2003;207: 15–21.
61 Mempel M, Voelcker V, Kollisch G, Plank C, Rad R, Gerhard M, Schnopp C, Fraunberger P, Malli AK, Ring J, Abeck D, Ollert M: Toll-like receptor expression in human keratinocytes: nuclear factor kappaB controlled gene activation by *Staphylococcus aureus* is toll-like receptor 2 but not toll-like receptor 4 or platelet activating factor receptor dependent. J Invest Dermatol 2003;121: 1389–1396.
62 Gavrin AJ, Gonyer RG, Blizard KG, Santos L: Medical textiles for uniform bacteria mitigation. 24th Army Sci Conf Proc, Orlando, 2004, KS-16. www.asc2004.com
63 BfR – Federal Institute for Risk Assessment: Exercise caution when using disinfectants! Press release 2004. www.bgvv.de
64 Lansdown AB, Williams A: How safe is silver in wound care? J Wound Care 2004;13:131–136.
65 Meloni GA, Schito GC: Microbial ecosystems as targets of antibiotic actions. J Chemother 1991;3(suppl 1):179–181.
66 Savey A, Fleurette J, Salle BL: An analysis of the microbial flora of premature neonates. J Hosp Infect 1992;21:275–289.
67 Korting HC, Lukacs A, Braun-Falco O: Mikrobielle Flora und Geruch der gesunden menschlichen Haut. Hautarzt 1988;39:564–568.
68 Takai K, Ohtsuka T, Suda Y, Nakao M, Yamamoto K, Matsuoka J, Hirai Y: Antibacterial properties of antimicrobial-finished textile products. Microbial Immunol 2002;46:75–81.
69 Chen D, Haviland-Jones J: Human olfactory communication of emotion. Percept Mot Skills 2000;91:771–781.
70 Rikowski A, Grammer K: Human body odour, symmetry and attractiveness. Proc R Soc Lond B Biol Sci 1999;266:869–874.
71 Austin C, Ellis J: Microbial pathways leading to steroidal malodour in the axilla. J Steroid Biochem Mol Biol 2003;87:105–110.
72 Kalyon BD, Olgun U: Antibacterial efficacy of triclosan-incorporated polymers. Am J Infect Control 2001;29:124–125.
73 Wollina U: Streit um antimikrobiell ausgerüstete Textilien – Geruchskiller in Socken und Höschen: Gefahr für die Haut? (Interview). MMW – Fortschr Med 2004;146:14.
74 Pirot F, Millet J, Kalia YN, Humbert P: In vitro study of percutaneous absorption, cutaneous bioavailability and bioequivalence of zinc and copper from five topical formulations. Skin Pharmacol 1996;9:259–269.
75 Moss T, Howes D, Williams FM: Percutaneous and dermal metabolism of triclosan (2,4,4′-trichloro-2′-hydroxydiphenyl ether). Food Chem Toxicol 2000;38:361–370.
76 Ohge H, Furne JK, Springfield J, Ringwala S, Levitt MD: Effectiveness of devices purported to reduce flatus odor. Am J Gastroenterol 2005;100:397–400.
77 Wollina U, Heide M, Müller-Litz W, Obenauf D, Ash J: Functional textiles in prevention of chronic wounds, wound healing and tissue engineering; in Elsner P, Hatch K, Wigger-Alberti W (eds): Textiles and the Skin. Curr Probl Dermatol. Basel, Karger, 2003, vol 31, pp 82–97.
78 Kurz J: Laundering in the prevention of skin infections; in Elsner P, Hatch K, Wigger-Alberti W (eds): Textiles and the Skin. Curr Probl Dermatol. Basel, Karger, 2003, vol 31, pp 64–81.

U. Wollina, MD
Head of Department
Department of Dermatology, Hospital Dresden-Friedrichstadt
Academic Teaching Hospital of the Technical University of Dresden
Friedrichstrasse 41
DE–01067 Dresden (Germany)
Tel. +49 351 480 1210, Fax +49 351 480 1219, E-Mail wollina-uw@khdf.de

Burg G (ed): Biofunctional Textiles and the Skin.
Curr Probl Dermatol. Basel, Karger, 2006, vol 33, pp 17–34

Silver in Health Care: Antimicrobial Effects and Safety in Use

Alan B.G. Lansdown

Imperial College Faculty of Medicine, Charing Cross Hospital, London, UK

Abstract

Silver has a long and intriguing history as an antibiotic in human health care. It has been developed for use in water purification, wound care, bone prostheses, reconstructive orthopaedic surgery, cardiac devices, catheters and surgical appliances. Advancing biotechnology has enabled incorporation of ionizable silver into fabrics for clinical use to reduce the risk of nosocomial infections and for personal hygiene. The antimicrobial action of silver or silver compounds is proportional to the bioactive silver ion (Ag^+) released and its availability to interact with bacterial or fungal cell membranes. Silver metal and inorganic silver compounds ionize in the presence of water, body fluids or tissue exudates. The silver ion is biologically active and readily interacts with proteins, amino acid residues, free anions and receptors on mammalian and eukaryotic cell membranes. Bacterial (and probably fungal) sensitivity to silver is genetically determined and relates to the levels of intracellular silver uptake and its ability to interact and irreversibly denature key enzyme systems. Silver exhibits low toxicity in the human body, and minimal risk is expected due to clinical exposure by inhalation, ingestion, dermal application or through the urological or haematogenous route. Chronic ingestion or inhalation of silver preparations (especially colloidal silver) can lead to deposition of silver metal/silver sulphide particles in the skin (argyria), eye (argyrosis) and other organs. These are not life-threatening conditions but cosmetically undesirable. Silver is absorbed into the human body and enters the systemic circulation as a protein complex to be eliminated by the liver and kidneys. Silver metabolism is modulated by induction and binding to metallothioneins. This complex mitigates the cellular toxicity of silver and contributes to tissue repair. Silver allergy is a known contra-indication for using silver in medical devices or antibiotic textiles.

Silver is a precious metal found in many parts of the world. The date of its discovery is not documented, but early manuscripts describe its medicinal properties and the value of silver vessels and coins in purifying the drinking water. Since these early days, silver has been used in a wide range of medical devices including bone prostheses, surgical sutures and needles, cardiac implants, catheters, dentistry, wound therapy and surgical textiles [1]. Whilst much of the early enthusiasm for using silver may have stemmed from its aesthetic value as a precious metal, evidence over the past 200 years increasingly points to its proven ability to protect the human body from infectious diseases. The French surgeon Credé [2] claimed that 0.5–1.0% silver nitrate reduced the incidence of neonatal eye infections in his clinic from 10.8 to about 2%, although details of this work are not available. It is anecdotal that whereas early 19th century surgeons like Credé and William Halstead chose silver foil and silver nitrate to protect wounds against disease, the actual isolation of infectious agents (bacteria) and their sensitivity to silver and other metals derive from the classical studies of Louis Pasteur and the postulates of infectious diseases by Robert Koch several years later [3].

Classical surgical studies demonstrating the antiseptic properties of silver date back to the times of Ambroise Paré (1517–1590), who used silver clips in facial reconstruction, and Halstead [4], chief surgeon of the Johns Hopkins Medical School, who employed silver wire sutures in surgery for hernia and found silver foil an effective means of controlling postoperative infections in surgical wounds. Silver nitrate has a long history in treating infectious diseases and proved an efficacious antiseptic for wound care for more than 150 years. Early clinical observations indicated that silver nitrate complexes with proteins in skin wounds to form 'resistant precipitates' and that the local antibacterial action can be easily controlled. This antiseptic action extends 'quite deeply' into wounds with silver forming soluble double salts of silver albuminates and silver chloride in the tissues. Its caustic and astringent properties are well documented in early pharmacopoeias when lunar caustic, silver nitrate pencils and strong silver preparations proved beneficial in eliminating calluses, warts and unsightly wound granulations [5]. Silver nitrate is still used in wound care and burns clinics today despite its astringency and ability to discolour the tissues.

Silver metal and silver nitrate formed the mainstay of antibiotics suitable for medicinal use up to the 1920s. Pharmaceutical studies of the 1920–1940 era provided fundamental knowledge on the antimicrobial action of silver. Von Naegeli (1895) is accredited with observing that silver exerts an 'oligodynamic' action on bacteria, namely that it exerts a lethal effect at very low concentrations. Later, Clarke [6] reported that bacteria, trypanosomes and yeasts are killed by silver at concentrations from 10^5 to 10^7 ions per cell, this concentration being equivalent to the 'estimated number of enzyme-protein molecules per cell'. More recent research suggests that most pathogenic organisms are killed in vitro at concentrations of 10–40 ppm Ag^+ with particularly sensitive organisms susceptible to 60 ppm [7].

In an attempt to overcome the irritancy of silver nitrate solutions, colloidal silver preparations were introduced into pharmacopoeias about 80 years ago [5]. Pharmacologists presumed that, by precipitating silver in the form of silver proteinate or colloidal solution, they could provide an efficacious antiseptic without undesirable side-effects. Although early colloidal silver products identified as mild or strong silver protein, colloidal silver halides and silver proteins achieved some popularity, they were superseded by newer and safer antiseptics, notably penicillins and silver sulphadiazine [8].

The introduction of silver sulphadiazine marked a renaissance in the use of silver in wound care. Whilst researching the antibiotic therapies available for controlling *Pseudomonas aeruginosa* in burn wounds, Fox [9] combined the antiseptic properties of silver with sulphonamide to provide a broader spectrum and safer antibiotic for use in burn wounds and surgery. Silver sulphadiazine and silver nitrate have been highly successful in controlling infections for many years even though the emergence of sulphonamide-resistant bacteria led to a temporary withdrawal of silver sulphadiazine in some hospitals in the mid-1970s [10]. Silver sulphadiazine has achieved a wider use in recent years with its inclusion in coatings for indwelling catheters and cardiac devices.

Improved technology permits manufacturers to enhance the delivery of silver ions to wounds to provide a safer and more efficacious antibacterial action (including methicillin-resistant *Staphylococcus aureus*, MRSA, and vancomycin-resistant enterococci) and effective prophylaxis against wound re-infection. The recent development of sustained silver release dressings marks a second renaissance of silver in wound therapy [8, 11]. The variety of silver release dressings now licensed in Europe and the USA differs greatly in composition, mechanism of presumed action and rates of silver release. They are variously tailored with recommendations for treating acute surgical wounds, burns, chronic or indolent wounds with profound exudation, unpleasant odours and severe patient discomfort.

Experience gained in the use of silver in wound care has inevitably led to the development of silver antibiotics in other medical devices. Thus, silver metal or a suitable silver compound is incorporated in polymers and resins used in the construction of medical devices, catheters, prostheses, bone cements etc., or it has been applied as an antibiotic 'coating' to silicone, textiles and other materials. Whilst clinicians have reported some success in reducing bacterial contamination and associated bacteraemias, recurrent problems have been encountered in biofilm transformation. These calcareous colonies of resistant bacterial and fungal infections form as a means of self-preservation. Biofilms are prevalent in indwelling catheters and implants, and are resistant to antibiotics and a host's own immune system. Although silver exhibits some ability to reduce bacterial adhesion as a preliminary step to biofilm formation, its success in eliminating or fully protecting against infections is limited.

The use of silver and silver release compounds in textile technology represents a new and exciting progression in health care [12]. Medicated clothing for nurses working in intensive-care clinics is a potentially beneficial means of controlling life-threatening nosocomial infections including MRSA as well as adding to levels of personal hygiene. The technology is discussed elsewhere in this publication, but antimicrobial efficacy against a range of pathogenic bacteria and fungi has been demonstrated in in vitro cultures. A feature of modern technology relevant to silver is compliance with safety standards. At the moment, most safety data and details on the metabolism and elimination of silver derive from occupational exposures complemented by clinical studies with silver sulphadiazine, but this holds clear clinical indications for the use of silver in other products.

Silver in Medical Devices and Textiles

Chemistry of Silver and Compounds Available for Antibiotic Action

Silver occurs naturally as two isotopes – Ag^{107} and Ag^{109} – in approximately similar proportions. It exhibits three oxidation states – Ag[1], Ag[2] and Ag[3] – but only compounds of the Ag[1] state are sufficiently stable to be of relevance as antibiotics in medical devices and textiles [7]. Like all other metals, silver is an electron-positive element with the Ag^+ cation showing a profound ability to interact with and bind proteins and anions in a medium. Additionally Ag^+ binds receptor groups on the surfaces of adjacent cells, bacteria and fungi/yeasts. Silver metal and the majority of silver compounds ionize in the presence of water, body fluids and tissue exudates to some extent to release Ag^+ or other 'biologically active silver ions' for antibiotic action or absorption into adjacent human tissues. The chemistry of silver is not well documented, and accurate data on relative ionization rates for the compounds commonly used in medical devices are not available (table 1).

Occasionally, documents fail to identify the nature of the chemical source of silver in products referring only to 'silver content' or 'ionic silver'. The ionizing capacity of the silver metal or silver compound is critical in comparing their antimicrobial activities and in predicting the possible toxicity or health risk. To be effective in killing pathogenic organisms, each silver source should release silver ions. The expressions 'activated' or 'hydro-activated' are used colloquially to denote the bioactive state of the silver ion. The silver cation binds strongly to electron donor groups of biological molecules containing sulphur (−SH), oxygen and nitrogen.

Table 1. Silver compounds used in medical devices and textiles

Compound	Ionizing capacity
Metallic silver (incl. nanocrystalline forms and silver coatings)	low ($<1\ g \cdot ml^{-1}$)
Phosphate	moderate
Nitrate	very high
Chloride	low
Sulphate	moderate
Zeolite	?
Sulphadiazine complex	high
Colloidal silver preparations	moderate to high
Allantoinates	?
Oxide (Ag_2O)	low

Metallic silver has been used in wound care products over many years even though it ionizes slowly. The development of nanochemistry has facilitated the production of microfine silver particles (<20 nm diameter) with greatly increased 'solubility' and release of silver ions (70–100 ppm). Ionization of silver metal is proportional to the surface area of the particle exposed. Ionization of silver or silver compounds is enhanced also by electric currents which are increasingly used in medical devices for orthopaedic surgery and wound care therapies [13].

Biological Properties of Silver

Silver is not a recognized trace metal but occurs in the human body at low concentrations ($<2.3\ \mu g \cdot l^{-1}$) due to ingestion with food or drinking water, inhalation and occupational exposures [7, 14]. Blood silver (argyraemia) is a measure of silver exposure from all sources. Clearly, occupational exposure or medicinal use of silver as an antibiotic in wound dressings, indwelling catheters, cardiac devices and in orthopaedic surgery will be associated with higher than normal blood levels and may be a safety concern [14]. Uptake of silver through mucous membranes (urethra) or by the haematogenous route from indwelling catheters is not well documented.

The percutaneous uptake of silver from medicated textiles like Sea Cell® [12] is not known. Medicated fabrics like cellulose fibres, rayon etc. will be in contact with the human skin for prolonged periods. Silver ions released in the presence of sweat, sebum and any moisture accumulate on the skin surface, and some will penetrate the superficial layers of the skin to precipitate as silver sulphide in the stratum corneum. Some will be bound by chloride ions in sweat but a minute proportion can be expected to penetrate into the circulation bound to albumins and other proteins [15, 16]. Hair and nail growth provides a route for the excretion of silver from the human body, but most will be eliminated via the liver and kidneys [14]. Hot weather and high humidities leading to hyperhidration will promote silver uptake through the skin and mucous membranes, but toxic risks are predictably low, except in individuals sensitized to silver.

Silver in Wound Care

Since early reports of clinicians using silver wire sutures, silver foil protection and silver nitrate in soft tissue surgery, silver has been the antibiotic of choice as a prophylactic or therapeutic against pathogenic infections in skin wounds, burns and transplant surgery [1, 7, 8]. As newer technology has come to hand, manufacturers of wound dressings have increasingly been able to tailor their products to suit wound type, infectious status and clinical features (pain, exudates, granulation tissue etc.). The amount of silver released can be controlled over the expected period of use of the dressings. Present approaches in wound dressings are towards safer and more efficacious dressings, cost-effectiveness and ease of use [17]. Basic principles of silver technology learned in the early days are equally relevant now.

Wound therapies employing silver as antibiotic range from silver metal, silver nitrate and silver sulphadiazine to the new generation of sustained silver release dressings. Emphasis in the silver release dressings is placed upon the nature of the silver source and patterns of release of silver ion. Even now there is a hot debate as to the clinical benefits of the fast high-concentration (bolus-like) silver ion release compared to the gentle, more sustained approach. Either way, wound care embracing silver antibiotic necessarily involves either (a) prophylaxis – to provide a barrier function in protecting acute skin damage (as in postoperative surgery) from nosocomial and idiopathic infections – or (b) therapeutics – to alleviate the microbiological burden in acute and chronic wounds.

The therapeutic management of chronic wounds increasingly observes principles of wound bed preparation [18], which can be critical in advancing the repair of chronic indolent wounds and ulcers. Wound bed preparation involves controlling the balance between commensal bacteria and pathogenic organisms.

Silver Nitrate

Silver nitrate is caustic and irritant at concentrations exceeding 1%; in contact with living tissue it can cause leakage of cellular electrolytes including sodium and potassium. However, it is an excellent antibacterial agent and at 0.5% is particularly effective in inhibiting *P. aeruginosa* which can prove fatal in burn wounds. Silver nitrate is claimed to be superior to many other antibiotics including chlorhexidine and silver sulphadiazine, especially in eliminating more resistant strains of *Streptococcus pyogenes, S. aureus* and *P. aeruginosa*

[10]. Silver nitrate compresses are claimed to reduce levels of infection in severe burns by up to 70% and significantly reduce mortality. It exhibits haemostatic properties and may be useful during minor surgery.

Silver Sulphadiazine

Silver sulphadiazine represents a second generation of silver antibiotics. Developed by Charles Fox in 1968, this complex combines the antibiotic properties of silver with a sulphonamide that proved invaluable in controlling wound infections in World War II [9]. It avoids many of the disadvantages of silver nitrate and at 1% in a cream base provides suitable prophylaxis for burns, chronic leg ulcers and pressure sores. Although silver sulphadiazine is sparingly soluble in water, it ionizes readily in body fluids to release silver ions. The amphiphilic (fat- and water-soluble) cream base enhances the penetration of silver sulphadiazine through intact skin and skin wounds allowing up to 10% to reach the systemic circulation [19]. Depending on the severity and depth of wounds, systemic silver sulphadiazine concentrations may reach up to 300 $\mu g \cdot l^{-1}$, absorption being higher where application is made to partial-thickness wounds with greater vascularity than full-thickness lesions. At concentrations of up to 50 $mg \cdot l^{-1}$, silver sulphadiazine is claimed to be effective against up to 95% of bacteria commonly found in skin wounds.

The low toxicity and antibacterial efficacy of sulphadiazine have been developed in a number of wound dressings using different vehicles and polymers (polyethylene glycol + poly-2-hydroxyethyl methacrylate, liposomes, poly-*L*-leucine and cadaver skin). Many have not progressed past the early laboratory stages, but a new lipocolloid formulation containing 3.75% silver sulphadiazine (Urgotul® SSD, Urgo-Parema Medical) is currently available and efficacious in treating acute and chronic wounds infected with a wide range of infections including MRSA [20]. Silver sulphadiazine cream 1% (Flamazine®) is still widely used either alone or in combination with other silver dressings. It is well tolerated by most patients and is preferred in burn clinics. A further development of particular benefit in burn wound therapy is Flammacerium® cream comprising 1% silver sulphadiazine with 2.2% cerium nitrate [21]. Both substances exhibit antibacterial properties, but the cerium ion actively reduces inflammatory changes, provides transitory immunosuppression against toxic factors released in tissue destruction and reduces mortality. Flammacerium is recommended in the treatment of severe burn wounds where excision is impractical and where infection is a serious health risk [21].

Sustained Silver Release Wound Dressings

Sustained silver release dressings presently available vary greatly in their technology, silver content, patterns of silver ion release and recommendations

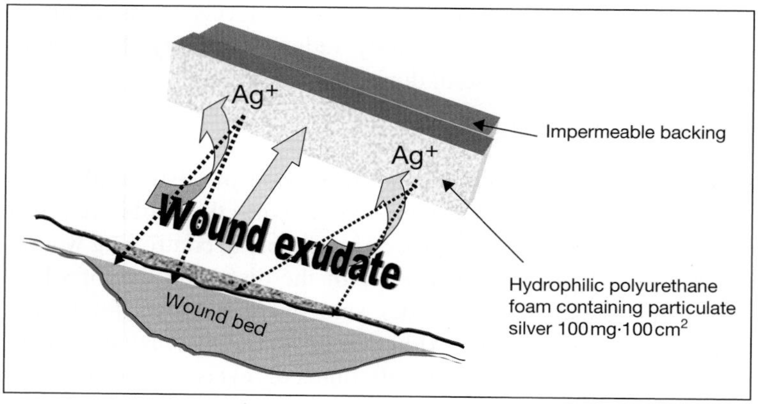

Fig. 1. Contreet® Foam Wound Dressing (Coloplast a/s, Humlebaek, Denmark). This dressing absorbs wound exudates into the hydrophilic matrix. Silver ions released within this matrix kill bacteria and fungal infections and inactivate toxins.

for clinical application [14, 20]. In each case, wound fluids and tissue exudates trigger the release of free silver ions for antimicrobial action or for absorption into tissues of the wound bed. Ideally, silver ion release will be sustained for the expected life-span of the dressing (up to 7 days). Three main forms of dressing are currently available:

(1) those releasing high levels of silver for rapid antimicrobial action;
(2) dressings that absorb wound exudates and where silver ions released provide sustained antimicrobial action (fig. 1);
(3) dressings that release silver sulphadiazine (Urgotul SSD).

Some dressings embody features of the first two of these considerations but all release more silver than the 10–40 ppm deemed necessary for appropriate antimicrobial action [14]. The excess silver compensates for that bound by scavenger anions or proteins in the wound bed (including wound debris) and which is unavailable for killing bacteria and fungi [20].

The silver content of dressings currently available varies from $<10\,mg\cdot 100\,cm^{-2}$ to more than $100\,mg\cdot 100\,cm^{-2}$. Several dressings provide a twofold objective of liberating silver as an antimicrobial agent and incorporating a material such as hydrocolloid, synthetic fabric or organic fibres to absorb exudates, odours and wound debris. These additional features are consistent with present clinical approaches to wound care, including wound bed preparation [18]. In each case, dressing materials or additives used are of low toxic risk and selected to absorb wound exudates, control pain and remove distasteful odours [7, 8, 17].

Nanotechnology has been used beneficially in the production of the high silver release dressing Acticoat (Smith & Nephew). Minute particles of silver contained within a 3-ply dressing of absorbent rayon polyester provide release of up to 70 ppm silver ions and antibacterial action lasting up to 3 days [14, 20]. This dressing is claimed to be effective against 150 wound pathogens including MRSA and more efficacious than silver nitrate or silver sulphadiazine. A further development of a wound dressing based on metallic silver employs activated-charcoal-impregnated cloth as a specific means of controlling wound odour attributable to pathogenic bacteria. The dressing (Actisorb Silver, Johnson & Johnson) is claimed to clear infections with the charcoal absorbing odours and bacterial toxins.

A third form of technology seen in wound dressings involves silver glass chemistry, similar to that developed for use in bone surgery and in orthopaedic prostheses. The polyurethane film dressings containing inorganic silver oxide, phosphate and polyphosphates have achieved notable success in the prophylaxis of acute wound therapy including open heart surgery. They protect wounds from nosocomial and other pathogenic flora whilst being entirely safe. The thin laminate slowly dissolves in wound fluids to release silver, calcium and phosphate ions for antibacterial action and wound repair.

It is unfortunate that although notable antibacterial and antifungal activity has been reported for all the sustained silver release dressings in in vitro experiments, clinical verification of their action is not generally available. Preliminary clinical trials in the Charing Cross Hospital have shown that patients treated with a range of dressings still show residual infections. It is unclear whether these bacteria are pathogenic or whether they are silver resistant [20]. So far, we have identified only 1 incidence of silver-resistant bacteria in a wound treated with a sustained silver release dressing (the coliform organism *Acinetobacter cloacae*) [Lansdown and Philip, unpubl.]. Further studies are now required to confirm whether this phenotypic silver resistance is a reflection of genetically determined changes.

Silver in Catheters

Clinical catheters for central vascular insertion or for urethral drainage are notoriously prone to infection with nosocomial organisms leading to biofilm formation. Indwelling central venous catheters are a major source of bacteraemias and candidaemias. Biofilm formation and accumulation of mineralized aggregates is a recurrent cause of catheter obstruction. New technology has been directed to engineer out risks of infection using silver as a component of catheter polymers or in the form of a hydrophilic coating to inhibit adhesion and colonization by pathogenic bacteria and yeasts with varying levels of success (fig. 2) [22, 23]. Catheters treated with silver metal (including nanocrystalline forms), silver oxide, silver sulphadiazine and other ionizable silver complexes have been evaluated in vitro, in animal experiments and in clinical

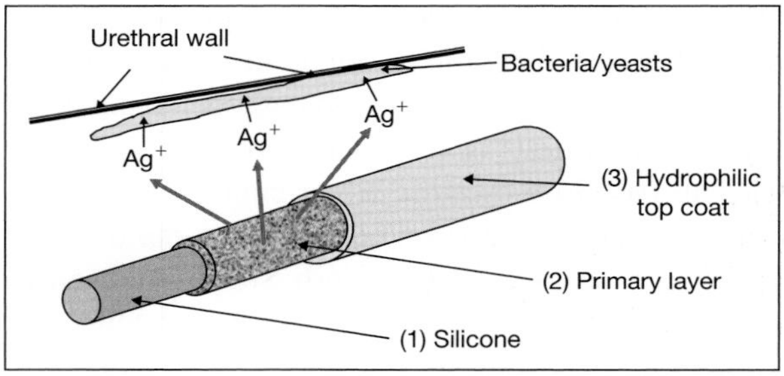

Fig. 2. Diagram of a silver-coated catheter (Dover®; Tyco Healthcare) showing a silicone core catheter with a primary coating of silver compound within a hydrophilic outer layer.

trials. Occasionally, other antibiotics including gentamycin, chlorhexidine gluconate or rifampicin have been included to complement the action of silver. Silver sulphadiazine and chlorhexidine gluconate act in a form of 'synergy' whereby the chlorhexidine gluconate serves to denature the bacterial cell membrane allowing improved ingression of silver ions. The copper ion exhibits a similar effect in medical catheters. Copper:silver filters are widely used in the purification of hospital water systems against *Legionella* sp. [8].

Silver-coated and silver-impregnated catheters have been evaluated for their antimicrobial action and capacity to prevent biofilm formation in a variety of in vitro systems, some simulating conditions of expected catheter use in clinical practice [23, 24]. Although laboratory experiments have substantiated the antimicrobial efficacy of silvered catheters, the observations have not been substantiated in live animal models or in clinical practice.

Silver has occasionally been employed to control infections associated with intraperitoneal catheters. Thus, silver as impregnate in cuffs, coatings or as ion-beam-assisted deposition has been applied as a coating for Silastic or other catheters but antibacterial protection has not been routinely confirmed [23].

Silver in Devices for Orthopaedic Surgery

Infection is a recurrent problem in association with external fixation pins and screws, prostheses, cements used in orthopaedic surgery and dental cavity fillings. Nosocomial organisms are documented causes of inflammatory changes, degenerative conditions, impaired healing and tissue function leading to lowered patient survival. Prophylactic concentrations of ionizable silver compounds have been included in these devices or applied in the form of a coating on the expectation that ions are released to control pathogenic infections

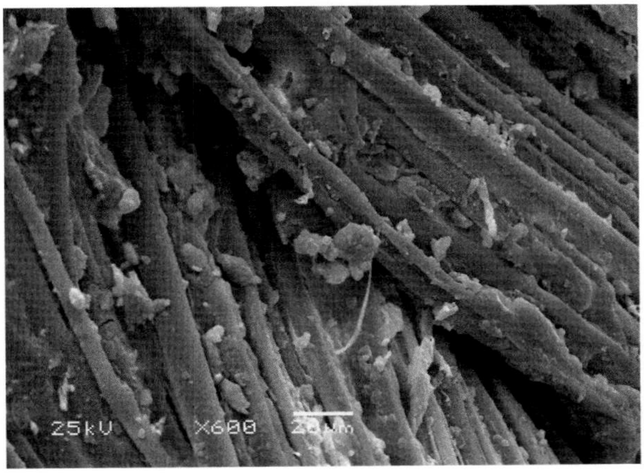

Fig. 3. Mersilk® (Ethicon) suture fibres following immersion in a slurry of silver-oxide-treated Bioglass granules. High-resolution SEM × 600.

without influencing repair mechanisms. At the moment, insufficient experimental or clinical evidence is available to justify the use of colloidal silver or other silver coating in fixation pins. Polymethacrylate bone cements laced with silver have proved effective in controlling *P. aeruginosa, S. aureus* and *Escherichia coli* in experimental studies but this has yet to be validated in clinical trials. The same is true where silver is used in bone prostheses.

The development of Bioglass® and its applications in bone surgery and in antiseptic sutures holds current interest in orthopaedic medicine. Bioglass biodegrades in the presence of tissue fluids and macrophages to release calcium, silicon and phosphate for bone repair and has clinical applications in hard and soft tissue repair. Silver oxide added to this bone cement has shown excellent antimicrobial potential in preclinical studies but approval for clinical studies is pending. In the same way, where medical silk sutures were 'doped' with silver oxide (fig. 3), antibacterial action as demonstrated in vitro and in experimental studies [Lansdown, Blaker and Boccaccini, unpubl.], clinical work is still pending.

Silver in Cardiovascular Surgery

Limited success has been achieved in using silver to control infections associated with cardiovascular devices including heart valve sewing rings, stents and prostheses. Although infections associated with prosthetic valve prostheses may lead to 80% fatality in clinical practice, devices such as the St. Jude Medical Silzone® sewing cuff, polyester-woven fabrics or knitted prostheses are not clinically acceptable or efficacious in controlling infections.

Darouiche [23] concluded that the validation of silver as an antimicrobial in cardiovascular devices was based on insufficient investigation and inadequate preclinical studies.

Silver in Textiles

The use of silver in medical textiles is at an early stage at present but holds many possible advantages in controlling clothing-borne nosocomial infections in hospitals and in personal hygiene products. It is expected to offer special advantages in protecting those people with acquired or inherited immunodeficiency conditions who are at greatest risk from even mildest infections. We have experience of silk sutures dipped in a slurry containing a Bioglass-silver oxide complex which have been developed at Imperial College (London) to be compatible with human cell lines in vitro. Electron microscopy has demonstrated distribution of the silvered granules on the surface of these silk sutures in a sufficiently robust form to withstand clinical incision at surgery (fig. 3).

In contrast, an antifungal and antibacterial cellulose fibre fabric (Sea Cell) has been developed at the Friedrich Schiller University in Jena (Germany) [12]. The natural cellulose fibres containing algal extracts provide a functional carrier for an ionizable silver compound which has shown commendable antimicrobial action against a range of pathogenic organisms in vitro at least. Clinical confirmation of this efficacy is urgently awaited.

Antibiotic Action of Silver

In his *Manual of Pharmacology* (1942) Sollemann [5] remarked that inorganic silver salts (especially nitrate) are astringent, caustic and antibiotic, that their action was readily controlled and their toxicity low. Since the 1940s when strong concentrations of silver nitrate as 'lunar caustic' or silver pencils were used to remove warts, unsightly granulations and skin calluses, silver has been a choice antibiotic (0.1–0.5%) for treating skin infections and bacterial and fungal infections associated with respiratory disease, bone and joint surgery, prostheses, cardiac devices, eye lesions and transplant surgery. Research has led to a greatly improved understanding of the mechanisms of antimicrobial action of silver and the molecular and genetic basis for silver resistance in bacteria and fungi [8, 20].

In its metallic form, silver is inert and exhibits no biocidal action. However, it ionizes in the presence of water or tissue fluids to release Ag^+ or other biologically active ions. This 'activated' ion shows a strong affinity for sulphydryl groups and protein residues on cell membranes. Importantly, silver exerts its antimicrobial action at low concentrations (1 ppm) and exhibits the so-called oligodynamic effect coined by von Naegeli in 1895. The lethal effect

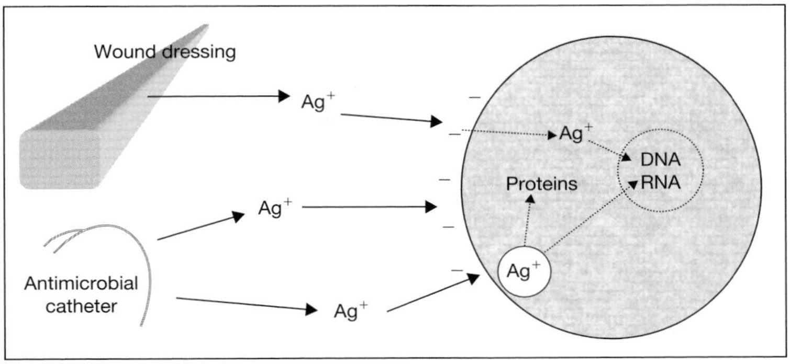

Fig. 4. Antimicrobial action of silver: (1) attachment to the bacterial cell membrane; (2) absorption/diffusion into the cell; (3) coagulation with bacterial proteins/enzymes.

in sensitive microflora was estimated to be equivalent to the number of intracellular enzyme systems to be inactivated, possibly in the range from 10^{-5} to 10^{-7}. Sensitive bacteria accumulate silver against a concentration gradient until lethality is reached [6].

More recently, microbiologists and molecular scientists have sought to unravel the genetic basis for resistance of bacteria to silver. Scanning and transmission electron microscopy has been used to examine:

- the action of silver on cell membranes;
- intracellular uptake and intracellular distribution of silver;
- interaction between silver and subcellular cytoplasmic components (enzymes, metal ion pathways etc.);
- the genetic and morphological basis for bacterial silver resistance.

Available evidence points to a direct correlation between bacterial lethality and 'available' concentrations of free silver in the medium. The silver ion that is chelated, bound or precipitated in insoluble complexes in tissue exudates or secretions is not available for antimicrobial action.

Silver is now known to inhibit a wide range of laboratory strains and type cultures of Gram-positive and Gram-negative strains. Broad extrapolations are made in predicting the responses of infections in the human body to silver. It is conceivable that where silver has been used in the components of materials used in medical devices or coatings, and where limited antimicrobial efficacy has been noted, a large proportion of silver ions released has been mopped up by chelating or binding agents in the micro-environment.

Mechanisms of antibacterial action by silver in sensitive organisms are complex and equivocal (fig. 4). Binding of silver to cell membranes and intracellular absorption is an obligatory first step; silver binds to electron donor receptors, notably disulphide, amino, imidazole, carbonyl and phosphate residues on membranes leading to intracellular absorption by endocytic vacuoles and phagocytosis. Inactivation of membrane-related enzymes like phosphomannose isomerase results in denaturation of the bacterial cell envelope and its functional capacity to regulate the inward diffusion of nutrients (e.g. phosphates, succinates) and limits the effusion of essential electrolytes and metabolites. Membrane damage has been identified by pitting and increased permeability as a prelude to lethality. The predominant intracellular effect of silver probably lies in its ability to impair key intracellular enzyme systems by impairing trace metals and electrolytes leading to defective respiratory pathways and RNA and DNA replication [7].

Silver resistance in bacteria at least has a molecular, morphological and genetic basis [8]. A silver-resistant strain of *E. coli* isolated from a burn wound containing two large plasmids failed to absorb or retain silver, whereas in a sensitive strain silver accumulation was fivefold higher. Genetic manipulation of these plasmids can dramatically alter silver uptake and hence the silver sensitivity/resistance of an organism. Electron microscopy and molecular techniques have shown that silver resistance encodes in a pericytoplasmic protein – *SilE* – and that this is expressed in the presence of metallothionein (a silver-induced, cystine-rich metal-binding protein). Cytoplasmic changes in a sensitive strain of *S. aureus* have been identified as electron-light regions in the cytoplasm associated with denatured DNA. Resistance (or R factor) can be transmitted through natural gene transfer mechanisms in bacteria, possibly similar to that seen in biofilm formation. Continued exposure of pathogenic infections in burns or other situations were associated with the clinical emergence of resistant strains in patients in the Birmingham Accident Hospital in the 1970s [10]. Experience suggests that emerging resistance among nosocomial and commensal bacteria may in part be controlled by restricting the use of antibiotics, changing practice in antibiotic therapies and using mixed antibiotics (e.g. silver + chlorhexidine) [22].

Biofilm formation is a major problem in the continued use of indwelling catheters and in orthopaedic materials [23, 24]. A wide variety of bacteria (and possibly yeasts and fungi) are known to adhere to surfaces of these devices, migrating along inner and outer surfaces to establish antibiotic-resistant colonies embedded in calcareous matrices. *Proteus mirabilis* has shown a remarkable ability to creep along such surfaces and develop biofilms. Offending bacteria undergo morphological and genetic changes in transforming from their natural 'pelagic' existence to the static way of life. Silver antibiotics in device coatings or materials have shown limited success so far in reducing the initial bacterial adhesion or in penetrating the biofilm matrix to achieve lethality. The reasons for this limited success are not immediately clear but may be related to the insufficiency of free silver ions released for antibacterial action in the devices. Much research is still needed to understand the mechanisms of bacterial resistance to silver in biofilm formation, why silver fails to penetrate the calcareous matrix and how silver interacts to influence natural human protective mechanisms. It is also unclear whether biofilms are likely to be a problem in the use of medicated textiles in nurses' uniforms or other hygiene products.

Metabolism and Toxicity of Silver in the Human Body

Silver is not a trace metal and serves no physiological role in the human body [7]. Silver absorbed into human tissues from antiseptic respiratory sprays, implanted medical devices, wound dressings or indwelling catheters can be expected to reach the systemic circulation, mostly as a protein complex. Argyraemias are more frequent in people exposed to silver occupationally or by environmental exposures including food and drinking water. Theoretically, silver can be deposited in any tissue in the human body but the skin, brain, liver, kidneys, eyes and bone marrow have received greatest attention [14]. Current information shows that the uptake and metabolism of silver are not well documented and are largely limited to a few clinical studies with silver sulphadiazine, where blood silver levels of $>300\,\mu g{\cdot}l^{-1}$ have been recorded [15, 16, 19]. Although silver released from intraurethral catheters is available for absorption through urinary tract membranes, the extent to which this occurs is not known. In these cases, primary attention is given to the influence of silver on bacteraemias and biofilm formation [24]. It is expected that as international regulatory requirements become more stringent, silver absorption data from a wide variety of products including textiles will become mandatory.

The silver ion is absorbed through the gastro-intestinal tract and through the lungs following inhalation of silver dust and vapour possibly by passive diffusion of a silver-protein complex or by protein transfer in a form of endocytosis [14]. Absorption through intact skin is low (<1 ppm?) since much of the free ion is precipitated as silver sulphide in outer parts of the stratum corneum. As much as 10% of silver sulphadiazine is absorbed through partial-thickness burns with exposure to high vasculature [19].

The intracellular metabolism of silver within human tissues is illustrated with reference to cells of a wound margin exposed to silver sulphadiazine. The silver ion absorbed into epidermal cells induces synthesis of the cystine-rich metal-binding proteins metallothionein 1 and 2. Silver avidly binds these proteins to form stable complexes [14, 20]. Increased cellular metallothionein favours the uptake of the key trace elements zinc and copper which in turn promote RNA and DNA synthesis, cell proliferation and tissue repair. Metallothionein thus provides a double action, it protects tissues against the potential toxic effects of a xenobiotic metal and promotes healing. It is expected that irrespective of the route of exposure, silver will bind to proteins in tissue fluids and exudates, particularly the albumins and globulins. These will be absorbed into the systemic circulation for distribution to soft and hard tissues. The rate of silver absorption is not known accurately for any route of exposure but silver in urine may provide a useful monitor for clinical or occupational exposure and absorption.

Silver absorbed into the human body accumulates in a transitory fashion in the liver, kidney, brain, lung and bone marrow with minimal or no toxic risk. Silver uptake by bone/teeth is low. On the other hand, silver absorbed intestinally or from antiseptic sprays does accumulate in the cornea of the eye (argyrosis) and in the dermis of the skin (argyria), especially tissues exposed to solar radiation. Argyrosis and argyria attributable to fine deposits of silver metal or silver sulphide tend to be long lasting or permanent but not life threatening. Argyria-like symptoms have been seen in patients subject to chronic silver sulphadiazine therapy or long-term exposure to dressings releasing high levels of silver [14]. These black silver sulphide deposits in wound debris are not in my opinion a true argyria; they are rarely permanent, not present in living tissues and are normally lost as wounds heal.

Silver or silver sulphide deposits in living tissues are rarely a cause for toxicological or physiological concern. In the liver and kidney which form the principal routes for silver excretion, silver granules accumulate in the cytoplasm of phagocytic cells, hepatocytes and renal tubular epithelium bound mostly in lysosomal vesicles. Silver is released into bile ducts and urinary ducts for excretion in the faeces and urine, respectively. Greater concern relates to the influence of chronic silver ingestion or silver sulphadiazine therapy on the bone marrow and circulating neutrophils [14]. Reports of neutropenia and erythema multiforme in burned children treated with silver sulphadiazine are now believed to be a self-limiting and transient change of no toxic significance. Leukopenia regressed when treatment was withheld. However, silver is known to sensitize predisposed individuals; this is a permanent effect and a contra-indication for using silver products for any infection. The incidence of silver allergy is not known (silver sensitivity tests are not routinely conducted except for diagnostic purposes). Clearly, evidence of silver allergy is a contra-indication for using any silver products therapeutically and avoiding silver-containing textiles which come into direct contact with the body.

Neurotoxic changes have been documented but not substantiated following the use of silver antibiotic in medical devices. Deposits of silver sulphide in the eye tend to be long lasting but pathological changes have not been identified in the tissues at concentrations as high as $970\,\mu g\ Ag \cdot g^{-1}$ tissue weight. Rarely, these silver sulphide deposits have been found to impair night vision but not optic function [14].

Inherent risks of using silver in wound care, medical devices or in textiles are predictably low. On the other hand, workers exposed to fine silver dust, silver vapours and ion beam technology in manufacturing processes are expected to be at greater risk to silver. The inherent risks of occupational silver exposure will be minimized by close adherence to standard procedures in good manufacturing practice, use of appropriate protective clothing and observation of stringent regulations such as Control of Substances Hazardous to Health (UK) which embodies a 'scientific assessment of risk'. Consumers exposed to silver in textiles are expected to be at minimal risk. Silver is not absorbed through intact skin, even in moist areas, to any great extent. It will be considerably lower than seen in patients exposed to 1% silver sulphadiazine cream (30% Ag) or wound dressings releasing at least 100 mg $Ag \cdot cm^{-2}$.

Discussion

Metallic silver and silver compounds are used widely in medical devices and health care products to provide antibacterial and antifungal action. Experience has shown that they are generally safe in use and effective in controlling pathogenic organisms. They do not achieve a 'germ-free' state in wounds, device-related infections or biofilm formation however. Biofilms are silver resistant, and silver-resistant bacteria have been isolated from burn wounds, chronic ulcers and nosocomial isolates [20]. It is expected that where silver has been unsuccessful in limiting infections, much of the free silver ion released for antimicrobial purposes has been mopped up by albumins, globulins, free anions and protein residues on cell membranes. At the moment, we do not know the minimal levels of silver ion necessary in any situation to clear infections, although recent research suggests that concentrations of free ion equivalent to 0.5–1.0 *M* silver nitrate will be adequate [Lansdown and Philip, unpubl.]. Manufacturers of new products should envisage providing a balance between the silver ion released for antibacterial purposes and the minimal toxic threshold. Although much initial research is still conducted in the laboratory with types of bacterial/fungal strains, further clinical and pharmacological studies are urgently required to examine the safety and efficacy of silver in human patients and volunteers with close attention to ethical considerations.

References

1 White RJ: An historical overview of the use of silver in wound management. J Care 2003;13:1–8.
2 Credé KSF: Die Verhütung der Augenentzündung der Neugeborenen, der häufigsten und wichtigsten Ursache der Blindheit. Berlin, Hirschwald, 1895.
3 Munch R: Robert Koch. Microbes Infect 2003;5:69–74.
4 Halstead WS: The operative treatment of hernia. Am J Med Sci 1895;110:13–17.
5 Sollemann T: A Manual of Pharmacology. Philadelphia, Saunders, 1942, pp 1102–1109.
6 Clarke AJ: General pharmacology; in Heffter A (ed): Handbuch der experimentellen Pharmacologie: Ergänzung. Springer, Berlin, 1937, p 4.
7 Lansdown ABG: A review of the use of silver in wound dressings: facts and fallacies. Br J Nurs 2004;13(suppl):S6–S13.
8 Lansdown ABG: Silver: its antibacterial properties and mechanism of action. J Wound Care 2002;11:125–131.
9 Fox CL: Silver sulphadiazine: a new topical therapy for *Pseudomonas aeruginosa* in burns. Arch Surg 1968;96:184–188.
10 Lowbury EJ: Problems of resistance in open wounds and burns; in Mouton RP, Brumfitt W, Hamilton Miller JMT (eds): The Rational Choice of Antibacterial Agents. Harrap Handbooks. London, Kluwer, 1977, pp 18–31.
11 Lansdown ABG: Silver in wound care and management; in Stephen-Haynes J (ed): Wound Care Society Educational Supplement. Wound Care Society, 2003, vol 1, No 3.
12 Hipler C, Elsner P, Fluhr JW: New silver-coated cellulose fibres: in vitro investigations into antibacterial and antimycotic activity. 2nd Eur Conf Textiles Skin, Stuttgart, 2004, No 5–1–29.
13 Becker RO, Spadaro JA: Treatment of orthopaedic infections with electrically generated silver ions: a preliminary report. J Bone Joint Surg 1978;60:871–881.
14 Lansdown ABG, Williams A: How safe is silver in wound care? J Wound Care 2004;13:131–136.
15 Wang XW, Wang NZ, Zhang OZ: Tissue deposition of silver following topical use of silver sulphadiazine in extensive burns. Burns Incl Therm Inj 1985;11:197–201.
16 Boosalis MG, McCall JT, Ahrenholz DH: Serum and urinary silver levels in thermal injury patients. Surgery 1987;101:40–43.
17 Lansdown ABG, Jensen K, Jensen MQ: Contreet foam and contreet hydrocolloid: an insight into two new silver-containing dressings. J Wound Care 2003;12:205–210.
18 Enoch S, Harding KG: Wound bed preparation: the science behind the removal of barriers to healing. Wounds 2003;15:213–229.
19 Coombs CJ, Wan AT, Masterton JP: Do burns patients have a silver lining? Burns Incl Therm Inj 1992;18:179–184.
20 Lansdown ABG, Williams A, Chandler S, Benfield S: Silver absorption and antibacterial efficacy of silver dressings. J Wound Care 2005;14:155–160.
21 Lansdown ABG, Myers S, Clarke J, O'Sullivan P: A reappraisal of the role of cerium in burn wound management. J Wound Care 2003;12:113–118.
22 Elliott T: Role of antimicrobial central venous catheters for the prevention of associated infections. J Antimicrob Chemother 1999;43:441–446.
23 Darouiche RO: Anti-infective efficacy of silver-coated medical prostheses. Clin Infect Dis 1999;29:1371–1377.
24 Saint S, Veenstra RH, Sullivan SD: The potential clinical and economic benefits of silver alloy urinary catheters in preventing urinary tract infection. Arch Intern Med 2000;160:2670–2673.

Dr. A.B.G. Lansdown
Honorary Senior Lecturer in Chemical Pathology
8, Fiddicroft Avenue
Banstead, Surrey SM7 3AD (UK)
E-Mail a.lansdown@imperial.ac.uk

Burg G (ed): Biofunctional Textiles and the Skin.
Curr Probl Dermatol. Basel, Karger, 2006, vol 33, pp 35–41

Antimicrobials and the Skin Physiological and Pathological Flora

Peter Elsner

Department of Dermatology, Friedrich Schiller University, Jena, Germany

Abstract

Healthy human skin is regularly colonized by nonpathogenic microorganisms. Bacterial genera isolated are coagulase-negative staphylococci and diphtheroid rods on the skin surface and propionibacteria in the infundibulum of the sebaceous glands. As for fungi, *Pityrosporum* (*Malassezia* spp.) is regularly present. The distribution and density of the flora is dependent on age and environmental factors such as sebum secretion, occlusion, temperature and humidity. Odor production in the axilla is related to the activity of aerobic diphtheroids. Antimicrobials may reduce the density of the skin resident flora, but they do not completely eliminate it. While antimicrobials may cause irritant and allergic contact dermatitis, no evidence exists that the use of antimicrobial substances may change the ecology of resident bacteria on the skin thereby leading to the overgrowth of pathogenic bacteria.

Antimicrobials have been used in textiles originally to prevent rotting, especially under adverse environmental conditions such as tropical climate. This dates back to World War II when cotton fabrics used in the South Pacific by the US Forces were protected from microbial decay.

The application of antimicrobial substances in textiles is now extended to textiles for medical uses (e.g. in the treatment of atopic dermatitis) but also in sports and leisure. In order to understand the use of antimicrobials in textiles, textile engineers should have some basic knowledge of the anatomy and physiology of the skin, especially its microbiology. While textiles should support physiological functions, they should not pose a risk to human health under normal or reasonably foreseeable use.

The Human Skin Flora

The skin is the largest organ of the human body. In the adult, it has a surface of 1.8 m^2 and a weight of 10 kg.

As the skin forms a barrier against harmful chemical and physical impact of the environment, it also protects the organism from infection by pathogens, parasites, fungi, bacteria and viruses.

One important protective factor against infection by pathogenic organisms is skin surface pH that physiologically is between 5 and 6 at nonoccluded sites [1]. Intertriginous areas such as the axillae, the toewebs, the genital and the perianal regions have a higher pH and are more prone to infection. In addition to pH, certain stratum corneum lipids have been shown to have inhibitory effects on pathogenic bacteria [2]. In recent years, defensins, peptides with antibiotic properties, have been demonstrated in human skin [3]. Peptidoglycan recognition proteins 3 and 4 are bactericidal against several pathogenic and nonpathogenic Gram-positive bacteria of transient, but not normal flora [4].

Despite these antibacterial substances present on the skin surface, the skin is not sterile but harbors nonpathogenic bacteria. The human skin flora is defined as the microbes present in healthy skin on the skin surface, within the stratum corneum, in the infundibulum of the sebaceous glands and in the hair follicles (fig. 1) [6]. It may be differentiated: the resident flora that is continuously present on the skin and thus may be regularly sampled, the transient flora that is only sampled at low frequency or density and the temporary resident flora that transiently grows on the skin without leading to infection.

The following genera make up the skin resident flora: micrococci with coagulase-negative staphylococci, *Peptococcus* spp., *Micrococcus* spp., diphtheroids with corynebacteria and *Brevibacterium* spp., propionibacteria and gram-negative rods.

The bacteria of the resident flora of the normal human skin live as microcolonies [7] on and between the layers of the stratum corneum [8] and in the follicles of the sebaceous glands [9]. In the uppermost layer of the stratum corneum, *Corynebacterium* spp. and sometimes microorganisms of the transient skin flora are present. In the infundibulum of the sebaceous glands, there is a topographic distribution of *Propionibacterium* spp. from the deepest part of the infrainfundibulum up to the entrance of the acroinfundibulum and of coagulase-negative staphylococci from the middle part of the infundibulum to its entrance.

Pityrosporum spp. are located near the ostium in the acroinfundibulum [9]. Skin bacteria are distributed within a three-dimensional space and not on a surface. Penetration of antimicrobial agents, especially into the infundibulum, is particularly difficult, because of the sebum/bacteria/corneal cell mass. Reduction

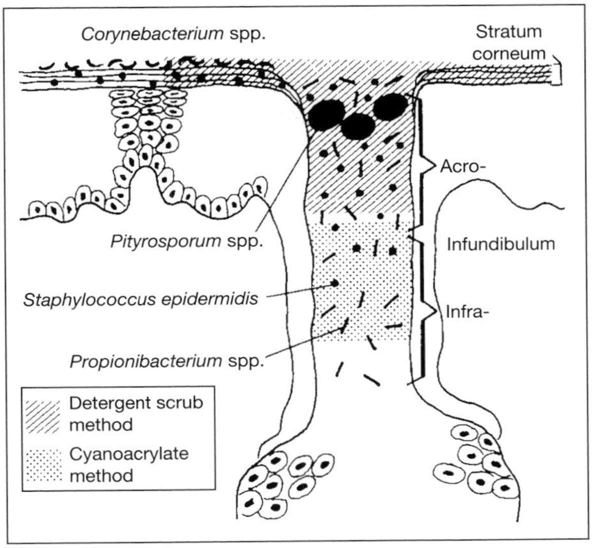

Fig. 1. Three-dimensional distribution of human physiological skin flora (based on Hartmann et al. [5]).

of the CFU counts per square centimeter of the bacteria of the resident flora to zero in vivo is impossible even after repeated applications of a potent skin disinfectant [10].

Korting et al. [11] have pointed out that the composition of the skin flora is subject to age: the streptococci which are found in infants disappear, and coryneform bacteria occur, which are mainly responsible for odor production. Anaerobic propionibacteria are more numerous in juveniles and young adults, a fact that may be explained by increased sebum production [11].

The relevance of skin resident flora for the healthy skin lies in the fact that it generates an ecological system protecting from pathogens. Thus *Staphylococcus epidermidis, Propionibacterium acnes,* corynebacteria and *Pityrosporum ovale* produce lipases and esterases that break triglycerides to free fatty acids leading to a lower skin surface pH and thereby unfavorable growing conditions for skin pathogens. *S. epidermidis* and *P. acnes* are known to produce antibiotics that may interfere with pathogenic organisms.

Endogenous and exogenous factors may influence the skin flora, thus facilitating skin infections. These factors include cellular and humoral immunity (e.g. diabetes, HIV infection), medication (e.g. Gram-negative folliculitis induced by long-term antibiotic treatment), environmental factors (e.g. humidity, temperature), and cosmetics and hygiene (e.g. antibacterial washes). Occlusive garments may be a risk factor for skin infections, and it is a well-known clinical experience that contaminated clothing may lead to a spread of infection, e.g. in staphylococcal folliculitis in hot, humid climates.

Skin Odor and Bacterial Flora

Already in the 1980s, several groups studied the relationship between skin odor and bacterial flora. Leyden et al. [12] showed that the axillary flora is a stable mixture of Micrococcaceae, aerobic diphtheroids and propionibacteria. Higher numbers of bacteria can be recovered from the axilla of persons with pungent axillary odor. This is related to the activity of aerobic diphtheroids, which could be ascertained in experiments with droplets of apocrine sweat placed on the forearm and inoculated with various bacteria: only diphtheroids generated the typical body odor. Cocci produced a sweaty odor attributable to isovaleric acid [12]. No ecological interactions between the members of the resident flora were found in the axilla [13]. Similar results were reported for foot odor, where high population densities of staphylococci and aerobic coryneform bacteria predispose to odor [14].

Malodorous androstenes (5α-androst-16-en-3-one, 4,16-androstadien-3-one, 5,16-androstadien-3β-ol, 5α-androst-16-en-3α-ol and 5α-androst-16-en-3β-ol) seem to be produced by bacteria in the axilla [15]. Recently, the biochemical pathways for the production of these steroid substances have been described and a metabolic map for axillary corynebacterial 16-androstene biotransformations was proposed, detailing potential enzyme activities [16].

Skin odor may be reduced by antibacterial agents without modifying the amount of sweat secreted. A clinical study has shown that the improvement of malodor correlates with a reduction of both micrococci (70%) or diphtheroids (73%) [17]. In persons presenting with persistent bromidrosis, the bacterial count per square centimeter did not significantly decrease and remained above 10^4 diphtheroids/cm^2 suggesting that body odor may be at least indirectly correlated to microbial counts with a bacteria threshold ranging around and above 10^4 [17].

Effects of Antimicrobials on the Bacterial Flora of the Skin

Antimicrobial substances are used to reduce the skin resident and transient flora, especially to eliminate pathogens in hygienic and surgical disinfection.

Table 1. Antimicrobial spectrum and characteristics of hand hygiene antiseptic agents (from Boyce and Pittet [19])

Group	Gram-positive bacteria	Gram-negative bacteria	Myco-bacteria	Fungi	Viruses	Speed of action	Comments
Alcohols	+++	+++	+++	+++	+++	fast	optimum 60–95%; no concentration persistent activity
Chlorhexidine (2% and 4% aqueous)	+++	++	+	+	+++	intermediate	persistent activity; rare allergic reactions
Iodine compounds	+++	+++	+++	++	+++	intermediate	causes skin burns; usually too irritating for hand hygiene
Iodophors	+++	+++	+	++	++	intermediate	less irritating than iodine; acceptance varies
Phenol derivatives	+++	+	+	+	+	intermediate	activity neutralized by nonionic surfactants
Triclosan	+++	++	+	–	+++	intermediate	acceptability on hands varies
Quaternary ammonium compounds	+	++	–	–	+	slow	used only in combination with alcohols; ecological concerns

+++ = Excellent; ++ = good, but does not include the entire bacterial spectrum; + = fair; – = no activity or not sufficient. Hexachlorophene is not included because it is no longer an accepted ingredient of hand disinfectants.

In Germany, the German Society for Hygiene and Microbiology has been publishing standards for the testing of disinfecting substances since 1959, and it publishes a list of tested and approved disinfectants including methods of hand decontamination and hygienic hand wash. The most widely used method for testing the efficacy of antiseptics in Europe is European standard EN 1500. In the USA, requirements for in vitro and in vivo testing of health care worker handwash products and surgical hand scrubs are outlined in the Food and Drug Administration's 'Tentative final monograph for healthcare antiseptic drug products' [18]. A list of the most frequently used approved topical antimicrobials with their spectrum of activity is given in table 1.

The main concern with the regular use of topical antimicrobial substances is the development of irritant and allergic contact dermatitis [19]. As a consequence of contact dermatitis in health care workers, the bacterial flora will change. Eczematous skin is characterized by papules, vesicles and, in irritant dermatitis, fissures that will lead to the exudation of intercellular fluid, thus promoting the growth of pathogens. In a prospective observational study of 40 nurses (20 with diagnosed hand irritation and 20 without), nurses with damaged hands did not have higher microbial counts ($p = 0.63$) but did have a greater number of colonizing species (means: 3.35 and 2.63, $p = 0.03$) [20]. Twenty percent of nurses with damaged hands were colonized with *Staphylococcus aureus* compared with none of the nurses with normal hands ($p = 0.11$). Nurses with damaged hands were also twice as likely to have Gram-negative bacteria ($p = 0.20$), enterococci ($p = 0.13$) and *Candida* ($p = 0.30$) present on the hands [20]. In using topical antimicrobial products, it is therefore important to select those with the lowest irritant and sensitizing potential under the given situation. When considering antimicrobials for the use in textiles with a possible transmission of the substances onto the skin, safety regarding irritation and sensitization should be shown by appropriate use tests employing noninvasive bioengineering technology [21]. A safety assessment as required by the European Cosmetics Directive is mandatory in these applications.

While evidence exists that the use of antimicrobial substances may change the ecology of resident bacteria of the gut thereby leading to the overgrowth of pathogenic bacteria, there are no such data for the skin [22]. As Jones [23] stated in 1999, based on current knowledge, the benefit from the use of topical antimicrobial wash products in combination with standard infection control and personal hygiene practices far outweighs the risk of increased antibiotic resistance.

References

1 Runeman B, Faergemann J, Larko O: Experimental *Candida albicans* lesions in healthy humans: dependence on skin pH. Acta Derm Venereol 2000;80:421–424.

2 Bibel DJ, Aly R, Shah S, Shinefield HR: Sphingosines: antimicrobial barriers of the skin. Acta Derm Venereol 1993;73:407–411.

3 Schroder JM: Epithelial antimicrobial peptides: innate local host response elements. Cell Mol Life Sci 1999;56:32–46.

4 Lu X, Wang M, Qi J, Wang H, Li X, Gupta D, Dziarski R: Peptidoglycan recognition proteins are a new class of human bactericidal proteins. J Biol Chem, in press.

5 Hartmann AA, Elsner P, Lutz W, Pucher M, Hackel H: Effect of the application of an anionic detergent combined with Fabry's tincture and its components on human skin resident flora. 1. Dermofug solution combined with either Fabry's tincture or 50 v/v% isopropanol. Zentralbl Bakteriol Mikrobiol Hyg B 1988;186:526–535.

6 Roth RR, James WD: Microbiology of the skin: resident flora, ecology, infection. J Am Acad Dermatol 1989;20:367–390.

7 Somerville DA, Noble WC: Microcolony size of microbes on human skin. J Med Microbiol 1973;6:323–328.
8 Montes LF, Wilborn WH: Location of bacterial skin flora. Br J Dermatol 1969;81(suppl 1):23.
9 Wolff HH, Plewig G, Januschke E: Ultrastruktur der Mikroflora in Follikeln und Komedonen. Hautarzt 1976;27:432–440.
10 Lilly HA, Lowbury EJ, Wilkins MD: Limits to progressive reduction of resident skin bacteria by disinfection. J Clin Pathol 1979;32:382–385.
11 Korting HC, Lukacs A, Braun-Falco O: Mikrobielle Flora und Geruch der gesunden menschlichen Haut. Hautarzt 1988;39:564–568.
12 Leyden JJ, McGinley KJ, Holzle E, Labows JN, Kligman AM: The microbiology of the human axilla and its relationship to axillary odor. J Invest Dermatol 1981;77:413–416.
13 Rennie PJ, Gower DB, Holland KT: In vitro and in-vivo studies of human axillary odour and the cutaneous microflora. Br J Dermatol 1991;124:596–602.
14 Marshall J, Holland KT, Gribbon EM: A comparative study of the cutaneous microflora of normal feet with low and high levels of odour. J Appl Bacterial 1988;65:61–68.
15 Nixon A, Mallet AI, Gower DB: Simultaneous quantification of five odorous steroids (16-androstenes) in the axillary hair of men. J Steroid Biochem 1988;29:505–510.
16 Austin C, Ellis J: Microbial pathways leading to steroidal malodour in the axilla. J Steroid Biochem Mol Biol 2003;87:105–110.
17 Guillet G, Zampetti A, Aballain-Colloc ML: Correlation between bacterial population and axillary and plantar bromidrosis: study of 30 patients. Eur J Dermatol 2000;10:41–42.
18 Food and Drug Administration: Tentative final monograph for healthcare antiseptic drug products: proposed rule. Fed Regist 1994;59:31441–31452.
19 Boyce JM, Pittet D: Guideline for hand hygiene in health-care settings. Recommendations of the Healthcare Infection Control Practices Advisory Committee and the HICPAC/SHEA/APIC/IDSA Hand Hygiene Task Force. Society for Healthcare Epidemiology of America/Association for Professionals in Infection Control/Infectious Diseases Society of America. MMWR Recomm Rep 2002;51:1–45, quiz CE1–4.
20 Larson EL, Hughes CA, Pyrek JD, Sparks SM, Cagatay EU, Bartkus JM: Changes in bacterial flora associated with skin damage on hands of health care personnel. Am J Infect Control 1998;26:513–521.
21 Rogiers V: EEMCO guidance for the assessment of transepidermal water loss in cosmetic sciences. Skin Pharmacol Appl Skin Physiol 2001;14:117–128.
22 Sullivan A, Edlund C, Nord CE: Effect of antimicrobial agents on the ecological balance of human microflora. Lancet Infect Dis 2001;1:101–114.
23 Jones RD: Bacterial resistance and topical antimicrobial wash products. Am J Infect Control 1999;27:351–363.

P. Elsner, MD
Department of Dermatology, Friedrich Schiller University
Erfurter Strasse 35
DE–07740 Jena (Germany)
Tel. +49 3541 937418, Fax +49 3541 937350, E-Mail elsner@derma.uni-jena.de

Burg G (ed): Biofunctional Textiles and the Skin.
Curr Probl Dermatol. Basel, Karger, 2006, vol 33, pp 42–50

Antimicrobial Textiles – Evaluation of Their Effectiveness and Safety

Dirk Höfer

Institute for Hygiene and Biotechnology, Hohenstein Research Center, Boennigheim, Germany

Abstract

The number of biofunctional textiles with an antimicrobial activity has increased considerably over the last few years. Whilst in the past it was predominantly technical textiles which had antimicrobial finishes, in particular to protect against fungi, nowadays textiles worn close to the body have been developed for a variety of different applications as far as medical and hygienic tasks. Together with the increase in new antimicrobial fibre technologies and possibilities in the hygienic and medical applications, the demand for proper test systems to evaluate the effectiveness as well as the safety of antimicrobial textiles rose. With the aid of agar diffusion and suspension tests, it is possible to record qualitative and quantitative data on the in vitro 'degree of effectiveness' of antimicrobial textiles. Test systems based on testing the biocompatibility of medical devices are suitable to evaluate the safety of antimicrobial textiles.

Applications and Objectives

The number of applications of textiles with antimicrobial activity has increased dramatically. A brief summary of the various fields of applications and associated products is shown by way of example in table 1.

From the data presented in table 1, one can deduce four main objectives behind the use of antimicrobial finishes [1]:

(a) to avoid the loss of performance properties as a result of microbial fibre degradation;
(b) to significantly limit the incidence of bacteria;
(c) to reduce the formation of odour as a result of the microbial degradation of perspiration;
(d) to avoid transfer and spread of pathogenic germs.

Table 1. Antimicrobial textiles and their fields of application

Medicine	Sport and leisure	Outdoor	Technology	Domestic
Support stockings	Shoes	Jackets	Wall hangings	Curtains
Antidecubitus mattress	Socks	Tents	Roof coverings	Coverings
Incontinence liners	T-shirts	Uniforms	Facade linings	Cloths
Encasings	Cycle wear	Personal protective	Air filters	Bath mats
Bedding filling	Team kit	Astro turf	Automotive	Sanitizers
Pillows	Jogging suits	Sunshades	Geotextiles	Underwear
Implants		Awnings		Carpets

Together with the increase in new antimicrobial fibre technologies and possibilities in the hygienic and medical applications [2], the demand for proper test systems to evaluate the effectiveness and safety of antimicrobial textiles rose [3].

Two Types of Activity

Before testing textiles with antimicrobial activities for their effectiveness, one has to take into account the technology and the bioactive substance which have been used to equip the textile. Three fundamental procedures have become established:

(a) the bioactive substance is applied directly to the spinning mass;
(b) the bioactive substance is applied to the fibre surface, e.g. with the aid of spacer molecules;
(c) the manufactured fabric is coated with a resinogenic finishing agent.

Depending on the type of technology and bioactive substance, textiles result with different antimicrobial active principles. Coarsely, these principles can be divided into two categories: materials with passive and materials with active effects [4].

Passive Antimicrobial Principles

Passive antimicrobial fibres do not contain any type of antimicrobial additives but have a quasi-antimicrobial effect, e.g. the microbial colonization of the textile is prevented by the surface structure of the fibre (microdomain-structured surfaces, lotus effect). In other words, the bacterial cells themselves are not affected – instead, the micro-organisms are prevented from adhering to the fibre surface (to prevent fibre degradation) or are attacked by non-leaching surface-active compounds. An example are so-called co-polyamines, antimicrobial polymers which have covalently attached polycationic compounds, that interact with the bacterial cell wall [5]. More recently, anti-adhesive, 'intelligent' polymers have been developed which prevent the formation of a biofilm on implants and thus offer a preventive measure for infections associated with implants (so-called plastic infections) [6].

Active Antimicrobial Principles

The majority of textiles with antimicrobial activity are materials with active finishes which contain specific active antimicrobial substances acting upon micro-organisms; either on the cell membrane, during metabolism or within the core substance (genome). A number of metallic compounds are also used; various silver compounds are currently very popular. Besides this, quaternary ammonium compounds, biguanides, amines and glucoprotamine displaying polycationic, porous and absorbent properties are active substances. Natural products such as chitosan are also being used more frequently [7]. The effectiveness of textiles with active antimicrobial activity is based on the so-called diffusion principle, i.e. the bioactive substance diffuses out at a variable rate from the finish or from within the fibre by ion exchange, by the substitution of cations from perspiration.

The antimicrobial principles listed above brought their own problems, however: it was often difficult or impossible to detect antimicrobial activity using conventional test methods, although the active substances themselves are clearly recognized and described as antimicrobial. This is due to the fact that a range of test systems only react to highly diffusive leaching substances (active principle) and therefore often give false-negative results where non-diffusive or weakly diffusive compounds are present.

Evaluating the Effectiveness of Antimicrobial Textiles

To record the quantitative reduction of bacteria by antimicrobial textiles, test systems have become established which specifically record this process. It is far beyond the scope of this article to compare all of the most important recognized international test methods for antimicrobial textiles (table 2). Therefore, we focused on the pros and cons of two major test systems: the agar diffusion test and the suspension test (challenge test).

Agar Diffusion Test

The agar diffusion test has a long-standing tradition in microbiology (e.g. testing the antibiotic resistance of germs). A probe is placed directly onto the surface of a germ-containing agar plate [8]. The test can easily be run with various micro-organisms. In textile research, this method can be recommended as a quick and preliminary qualitative method to distinguish between active

Table 2. Overview of the most popular standards to test the performance of textiles, fibres, yarns and polymers for antimicrobial effectiveness

Designation	Title	Principle
SN 195920-1992	Textile fabrics: determination of the antibacterial activity	Agar diffusion test
SN 195921-1992	Textile fabrics: determination of the antimycotic activity	Agar diffusion test
EN 14119:2003-12	Textiles-evaluation of the action of microfungi	Agar diffusion test
ASTM E 2149-01	Standard test method for determining the antimicrobial activity of immobilized antimicrobial agents under dynamic contact conditions	Challenge test
JIS Z 2801	Antimicrobial products – test for antimicrobial activity and efficacy	Challenge test
JIS L 1902-2002	Testing for antibacterial activity and efficacy on textile products	Challenge test

(i.e. clear zone of growth inhibition around the probe) and passive antimicrobial principles (no zone of inhibition). The diameter of the zone of inhibition may give an indication of the dimension of the antimicrobial activity or the release rate of the diffusive compound. Thus, it is not more than a semi-quantitative analysis. Nevertheless, the agar diffusion test is also favourable when testing textiles with passive antifungal activity (fig. 1), since fungi do not grow or are impaired in their growth on the specific probes. For textile materials other than fabrics, only very poor or imprecise conclusions can be drawn using this method.

Suspension Tests

With the aid of suspension tests such as the JIS 1902-2002 it is possible to record the maximum achievable in vitro 'degree of effectiveness' of the finished textiles as growth inhibitors of test bacteria [9]. This standard was not designed as a wear test for textiles, but it is an excellent method to assess the degree of the antimicrobial activity of textiles. Suspension tests can also be used to differentiate between a specific or general antimicrobial activity, and they can be modified to check the activity of textiles with the passive principle. When calculating and evaluating the antimicrobial activity, it is advantageous that the composition of the control and sample materials is identical, as in the test for specific antibacterial activity. The only difference must be the antibacterial component of the sample. This is the only way to guarantee that the test results reflect the specific antibacterial activity of the treatment over the test period.

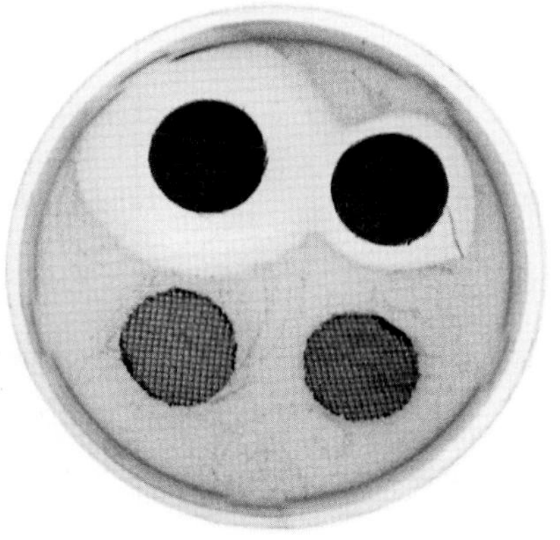

Fig. 1. Agar diffusion test with antimicrobial textiles. The two upper samples on the plate show a clear zone of inhibition (bacteria have been destroyed around the samples). This can be taken as an active antimicrobial principle. In contrast to this, the two lower fabrics do not show any signs of an active zone, although these probes are also treated with an antimicrobial finish (non-diffusive, passive antimicrobial principle).

Common Procedure: Test for Specific Antibacterial Activity

It is favourable to start with a log culture of bacteria (e.g. *Staphylococcus aureus* 6538 ATCC, *Klebsiella pneumoniae* DSM 789) diluted with sodium chloride to a definite suspension with 10^5 CFU (colony-forming units). In this dilution stage, the inoculum is deprived of virtually all reserve nutrients, in order to minimize factors influencing growth and directly record the specific effect of the antibacterial treatment applied to the textile fibres. It is also of advantage to conduct the suspension test with a sample material that has been given an antibacterial treatment and a reference material. The latter is a fabric of the same structure and chemical/physical properties and composition, but without the antibacterial treatment. Both materials are inoculated and incubated for 18 h at 36°C. The average value is taken from parallel tests (e.g. triple control). After the incubation, both materials are washed out with physiological salt solution with detergent supplement, and the number of separated bacteria is recorded in terms of CFU, e.g. using a spiral plater procedure. The bacterial growth over 18 h is calculated for the reference material and the sample material (CFU sample), respectively.

Calculation and Assessment

The reduction value in the growth of micro-organisms R is calculated, which expresses the difference in bacterial growth between the sample material and the reference material RM over 18 h. It is expressed in log stages as the bacterial growth-inhibiting, antibacterial activity:

Table 3. Assessment of the antimicrobial activity

Antibacterial activity	Specific antibacterial activity
Slight	0 to <1*
Significant	≥ 1 to <3
Strong	≥ 3

*Because of the instability of the bacterial growth, the biological variation (laboratory standard $\pm$ 0.5 log stages) should be taken into account in the assesment criteria, especially in the lower range/where the effect is slight. When bacteria are used, the term 'antimicrobial' can be specified to 'antibacterial'.

$$\text{specific antibacterial activity} = \log_{10} (\text{CFU}_{\text{RM/18 h}}) - \log_{10} (\text{CFU}_{\text{sample/18 h}}) = \text{R}$$
$$\text{specific antibacterial activity} = \text{BG}_{\text{RM}} - \text{BG}_{\text{sample}}$$

where BG = bacterial growth.

The $\log_{10}$ values for the specific antibacterial activity can be rated according to the assessment described in table 3.

When nutrients are limited during the incubation phase, depending on the fibre type changes may be observed in the bacterial population: on substrates/fibre types containing no nutrients, stagnation occurs, while natural fibres which contain nutrients (wool, silk) may promote growth. Occasionally a slight reduction in the bacterial culture may even be detected. However, by directly comparing a reference material with the treated sample, it is always possible to record the direct effect of the additional antibacterial treatment, because external factors (e.g. supply of nutrients) can largely be excluded and, because of the characteristics of the sample and reference materials, it can be assumed that any potential growth pattern will be the same.

Controls

To check the capacity for growth of the test bacteria, an internal growth control test should always be carried out using a defined test material. In experiments, values between $\pm$0.5 log stages can been achieved. To check the recovery rate, the 0-hour value should also be calculated. The bacterial suspension is also plated at 0 and 18 h and the CFU are counted.

Common Procedure: Test for General Antibacterial Activity

In the absence of a reference material with the same chemical/physical properties, the internal growth control material can be used as a reference value. In this case, the assessment should be specified to the term 'general antibacterial activity of textile' in order to distinguish clearly from the term 'specific antibacterial activity'. If there is no reference material, the sample and reference cannot be compared and evaluated purely with regard to the antibacterial treatment, i.e. directly. Additional treatment of the fibres in finishing processes, dyeing, washing etc. can affect the growth of bacteria and so contribute to the general antibacterial activity.

By comparing the 18-hour value for the internal growth control material with the 18-hour value for the sample, the total activity of the treated textile can be expressed using the following formulas:

$$\text{general antibacterial activity} = \log_{10}(\text{CFU}_{\text{IGM/18 h}}) - \log_{10}(\text{CFU}_{\text{sample/18 h}})$$
$$\text{general antibacterial activity} = \text{BG}_{\text{IGM}} - \text{BG}_{\text{sample}}$$

where BG = bacterial growth and IGM = internal growth control material.

Evaluating the Safety of Antimicrobial Textiles

More and more antibacterial textiles aim in particular at clients from hygiene-sensitive sectors such as the health sector and the foodstuffs industry. The fact that the clothing worn by doctors and nursing staff, as well as that worn by workers in food-processing companies, can play a critical role in the transfer of dangerous pathogens is indisputable [10, 11]. This is particularly true if the effectiveness has been proven beyond all doubt, i.e. on the basis of practical tests carried out by a neutral body. On the other hand, the term 'antibacterial' inevitably raises questions amongst wearers as to the skin compatibility of antimicrobial textiles worn next to the skin. Compelling legal reasons such as § 30 of the Ordinance on Foodstuffs and Articles for Domestic Use [12] require the use of innovative textiles to ensure there are no risks to health. The user of the textile products can therefore expect them to be safe in use and not to pose a health risk. If, for instance, an employer demands that specific workwear is worn or actually provides such workwear, he must ensure that this is safe as part of his duty of care towards his employee. For this reason, validated test methods have been developed to objectively assess the biological safety of textiles with antimicrobial activity on a scientific basis. The risk analysis is central to this test, in other words whether and in what way it is possible to assess the biological effects of these types of textile finishes on humans, what benefits the materials offer the wearer and whether these benefits are achieved without posing additional risks for the wearer.

Biological Safety Tests on Skin Test Systems

The basis for the biological safety tests for textiles is the EN ISO 10993 [13] for the biological evaluation of medical devices. This stipulates, depending

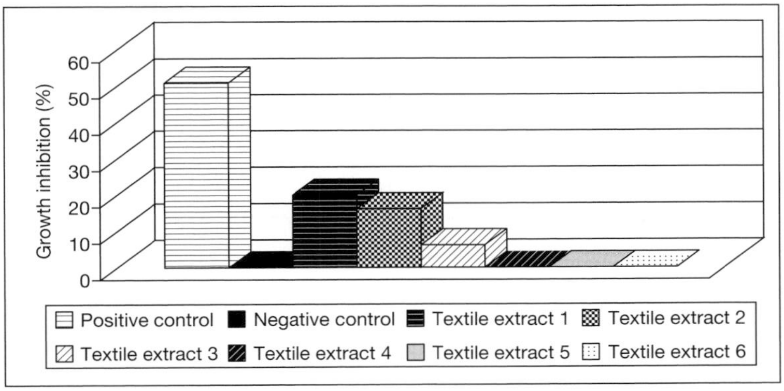

Fig. 2. Textile test for biocompatibility: dilutions of a textile extract (1–6) do not lead to a significant reduction in cell growth of fibroblasts. Only the positive control results in a tremendous and significant reduction in cell proliferation (>30%), indicating severe cytotoxicity.

on the type and duration of body contact, what risk analyses need to be carried out and which test methods need to be employed for this. The cytotoxicity (tissue compatibility) and sensitization and irritation potential are tested [14].

Cytotoxicity

When considering the tissue compatibility of textiles with antimicrobial finishes, it is particularly important to look at whether potentially cell-toxic substances (i.e. bioactive leaching substances) could be released from the material during normal wear. In the cytotoxicity test in accordance with EN ISO 10993, an extract from the textile is prepared using an artificial perspiration solution. The effects of this on the L 929 fibroblasts and HaCaT keratinocytes from the human epidermis provide information on potentially cell-toxic components. Figure 2 shows the results of a test for cytotoxicity, carried out on a antimicrobial active textile which was finished with silver and is used to treat cases of atopic dermatitis [15, 16].

Irritation

A classic test to determine the irritation potential of a substance is the Draize test, in which the substance to be tested is dripped onto the conjunctiva of a laboratory animal's eye in order to identify any potential irritants. The scientifically recognized hen's egg test on the chorio-allantoic membrane is an alternative to the animal test [17]. This has been validated among others by the European Centre for the Validation of Alternative Methods. It is possible to determine the irritation potential of substances which could be released from the textile material just as accurately by observing the blood vessels of the treated egg as using the animal test. In addition to the cytotoxicity test, the hen's egg test on the chorio-allantoic membrane therefore offers decisive additional security regarding the use of antimicrobial textiles. In conclusion, new biological test systems make it possible to scientifically determine the interactions between textiles and the skin accurately and to recognize and evaluate potential benefits and risks. The methods can be used as safety tests for textiles with antimicrobial activity.

References

1 Rouette H-K: Lexikon für Textilveredlung. Dülmen, Laumann, 1995.
2 Vigo TL: Antimicrobial polymers and fibres; in Seymour RB, Porter RS (eds): Manmade Fibres: Their Origin and Development. London, Elsevier Applied Sciences, 1993, p 214.
3 Wulf A, Moll I: Silberbeschichtete Textilien – eine ergänzende Therapie bei dermatologischen Erkrankungen. Akt Dermatol 2004;30:28–29.
4 Mucha H, Hoefer D, Assfalg S, Swerev M: Antimicrobial finishes, modifications, regulations and evaluation. Melliand 2002;4:238–243.
5 Thölmann D, Kossmann B, Sosna F: Polymers with antimicrobial properties. Eur Coat J 2003;1/2: 105–108.
6 Tiller JC, Liao CJ, Lewis K, Klibanov AM: Designing surfaces that kill bacteria on contact. Proc Natl Acad Sci USA 2001;98:5981–5985.
7 Takai K, Ohtsuka T, Senda Y, Nakao M, Yamamoto K, Matsuoka J, Hirai Y: Antibacterial properties of antimicrobial-finished textile products. Microbiol Immunol 2002;46:75–81.
8 Infante MR, Diz M, Manresa A, Pinazo A, Erra P: Note: microbial resistance of wool fabric treated with bis-Quats compounds. J Appl Bacteriol 1996;81:212–216.
9 JIS L 1902: Testing for antibacterial activity and efficacy on textile products, 2002.
10 Panknin G, Geldner G: Problemerreger auf Intensivpflegestationen unter Berücksichtigung von MRSA. Hartmann Wundforum 2000;4:15–21.
11 Sander CS, Elsner P: Pilzinfektionen und Textilien. Akt Dermatol 2004;30:18–22.
12 Ordinance on Foodstuffs and Articles for Domestic Use: § 30: Prohibition for the protection health Bundesgesetz blatt I.6. Sep. 2005.
13 International Standards Organization: EN ISO 10993: Biological evaluation of medical devices. Geneva, International Organization for Standardization, Part 1: Evaluation and testing, 1995.
14 Höfer D: Produktprüfung von Biofunktionsbekleidung. Melliand 2004;10:805–808.
15 Gauger A, Mempel M, Schekatz A, Schafer T, Ring J, Abeck D: Silver-coated textiles reduce *Staphylococcus aureus* colonization in patients with atopic eczema. Dermatology 2003;207: 15–21.
16 Abeck D, Mempel M: *Staphylococcus aureus* colonization in atopic dermatitis and its therapeutic implications. Br J Dermatol 1988;139:13–16.
17 Spielmann H: HET-CAM test. Methods Mol Biol 1995;43:199–204.

Dr. Dirk Höfer
Institute for Hygiene and Biotechnology, Hohenstein Research Center
DE–74357 Boennigheim (Germany)
Tel. +49 07357 271432, Fax +49 07357 271 91432, E-Mail d.hoefer@hohenstein.de

Burg G (ed): Biofunctional Textiles and the Skin.
Curr Probl Dermatol. Basel, Karger, 2006, vol 33, pp 51–66

Physiological Comfort of Biofunctional Textiles

Volkmar T. Bartels

Department of Clothing Physiology, Hohenstein Institutes, Boennigheim, Germany

Abstract

Statistics show that the wear comfort is the most important property of clothing demanded by users and consumers. Hence, biofunctional textiles only have a high market potential, if they are comfortable. In this work it is shown how the thermophysiological and skin sensorial wear comfort of biofunctional textiles can be measured effectively by means of the Skin Model and skin sensorial test apparatus. From these measurements, wear comfort votes can be predicted, assessing a textile's wear comfort in practice. These wear comfort votes match exactly the subjective perceptions of test persons. As a result validated by wearer trials with human test subjects, biofunctional textiles can offer the same good wear comfort as classical, non-biofunctional materials. On the other hand, some of the biofunctional treatments lead to a perceivably poorer wear comfort. In particular, the skin sensorial comfort is negatively affected by hydrophobic, smooth (flat) surfaces that easily cling to sweat-wetted skin, or which tend to make textiles stiffer. As guidelines for the improvement of the thermophysiological or skin sensorial wear comfort, it is recommended to use hydrophilic treatments in a suitable concentration and spun yarns instead of filaments.

There is no common definition of the term 'biofunctional textiles' in the literature, but they comprise materials which are e.g. antimicrobial or fungicidal [1] due to special finishes or fibre modifications, as well as textiles with cyclodextrin finishes [2], which act as 'cage molecules' and can enclose malodorous substances like sweat components or deliver drugs to the skin. Biofunctional textiles can be used in many different textile applications. Here, the focus will be on clothing:

- In the field of sportswear or occupational clothing, today biofunctional textiles are frequently used to suppress the build-up of sweat odours.

- Another important application for biofunctional textiles is the field of medical textiles [3]. As examples, antimicrobial textiles are intended to be used in work wear for hospital staff to improve hygiene by a higher protection against germ penetration and a lower risk of spreading germs from one place to the other. Another example is clothing for people suffering from neurodermatitis.

On the other hand, the wear comfort of clothing is a main quality criterion. The wear comfort does not only affect the well-being of the wearer, but also his performance and efficiency. Hence, it is appropriate to designate the wear comfort as the 'physiological function' of clothing. Its importance shall be demonstrated with the example of a surgeon working in an OR gown with only poor breathability [4]: after a certain tolerance time, the surgeon will start to suffer from heat stress, which impedes his mental performance. As a consequence, the risk of professional blunders increases, i.e. the poor physiological function of the OR gown endangers the health or even life of the patient!

Wear comfort is also a major sales aspect: according to the journal *World Sports Activewear,* 'in fact, comfort is the most important thing in clothing…, and it is coming from sportswear where consumers have become accustomed to the comfort' [5]. 94% of consumers would like their clothing to be comfortable, i.e. wear comfort is No. 1 in consumer expectations [6]. Consequently, in a survey 98% of specialized German dealers believe wear comfort to be an important or very important property of clothing [7, 8]. Hence, biofunctional textiles only have a high market potential, if their wear comfort is as good as that of normal, non-biofunctional materials.

But just recently, the influence of different biofunctional treatments of textiles on their wear comfort has been surveyed for the first time [9]. In that research project, several constructions for work wear, leisure wear and sports textiles were investigated. The data presented here are taken from a technical report [9] (in German), which is available from the author.

Aspects of Wear Comfort

After having recognized the importance of the wear comfort and the physiological function of clothing, one should define in more detail what wear comfort is about. In fact, wear comfort is a complex phenomenon, but in general it can be divided into 4 different main aspects [10]:

(1) The first aspect is denoted as the *thermophysiological wear comfort*, as it directly influences man's thermoregulation. It comprises heat and moisture transport processes through the clothing. Key words are thermal insulation, breathability, moisture management etc.

(2) The *skin sensorial wear comfort* characterizes the mechanical sensations, which a textile causes in direct contact with the skin. These perceptions may be pleasant like smoothness or softness, but they may also be unpleasant, if a textile is scratchy, too stiff or clinging to sweat-wetted skin. Textiles with poor skin sensorial wear comfort may even lead to mechanically induced skin irritations.

(3) The *ergonomic wear comfort* deals with the fit of the clothing and the freedom of movement it allows. The ergonomic wear comfort is mainly dependent on the garment's pattern and the elasticity of the materials.

(4) Last but not least, the *psychological wear comfort* is of importance. It is affected by fashion, personal preferences, ideology etc. The psychological aspect should not be undervalued: who would feel comfortable in clothing of a colour he or she dislikes?

Here, the thermophysiological and the skin sensorial wear comfort of biofunctional textiles are investigated, as they may both be influenced by the biofunctional treatment.

Measurement of Physiological Comfort

Wear Comfort as a Measurable Quantity

Many people believe that comfort is something individual to each person, which cannot be quantified or measured. But in fact wear comfort is directly related to physiological processes within our bodies.

For instance, the thermophysiological comfort is based on the principle of energy conservation. All the energy, which is produced within the body by metabolism M, has to be dissipated in exactly the same amount from the body [10, 11]:

$$M - P_{ex} = H_{res} + H_c + H_e + \Delta S/\Delta t$$

with P_{ex} the external work, H_{res} the respiratory heat loss because of breathing, H_c the dry heat flux comprising radiation, conduction and convection, and, last but not least, the evaporative heat flow H_e caused by sweating. If more energy is produced than dissipated, the body suffers from hyperthermia. On the other hand, too high a heat loss implicates hypothermia. Both lead to a change in the body's energy content ΔS with time Δt. ΔS may be either positive (leading to hyperthermia) or negative (hypothermia) and is zero for steady state.

As we have seen, the wear comfort is directly related to physiological processes. It is therefore accessible to a quantitative measurement. The testing techniques applied within the research project [9] are described in the following.

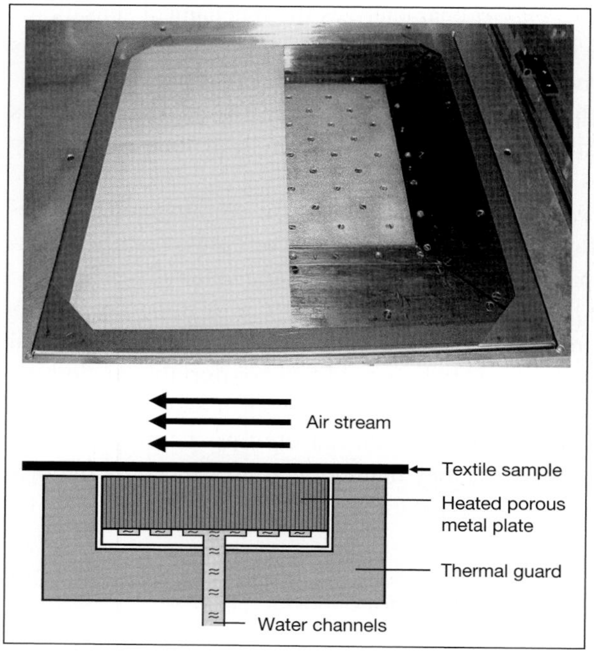

Fig. 1. Photo and schematic drawing of the Skin Model.

Skin Model

The thermophysiological wear comfort of textile materials is measured by means of the so-called Skin Model. The Skin Model is a thermoregulatory model of the human skin [12]. It is internationally standardized [13]. For protective clothing, it is the only test method for breathability, which is accepted within the European standardization.

A photo and a schematic drawing of the Skin Model are given in figure 1. The shown measuring unit is made of sintered stainless steel. Water, which is supplied by channels beneath the measuring unit, can evaporate through the numerous pores of the plate, just like sweat out of the pores of the skin. Additionally, the measuring unit is kept at a temperature of 35°C. Thus, heat and moisture transport are comparable to those of the human skin.

With the Skin Model different wear situations can be simulated [14]. Therefore, textile-specific parameters are obtained, characterizing the wear comfort in the different scenarios:

(1) *Normal wear situations* [13] are characterized by an insensible perspiration, i.e. the wearer does not recognize that he is sweating. Nevertheless, at least

30 g/h of water vapour is evaporated through the skin, which, in this case, acts as a semi-permeable membrane. The water vapour has to be transported through the textile by diffusion [15], i.e. the material has to be breathable. The textile-specific parameters characterizing normal wear situations are:

- the thermal insulation R_{ct}, which should be adjusted to climate and activity;
- the water vapour resistance R_{et}, which should be as low as possible and which corresponds to a high 'breathability';
- the water vapour permeability index

  $i_{mt} = 60\ (Pa/K)\ R_{ct}/R_{et}$

  which judges the breathability of fabrics taking their thermal insulation into consideration; it ranges between 0 (no breathability at all) and 1 (value of still air);
- the short-time water vapour absorbency F_i.

(2) With *heavier sweating*, e.g. when walking upstairs, the wearer already recognizes that he starts to sweat, but he is not sweat wetted yet. In these situations, the skin produces vaporous sweat impulses, which can be simulated with the Skin Model as well [16]. The textile has to maintain the microclimate humidity as dry as possible. Therefore, the material's abilities to transport and take up water vapour are decisive. Both are combined in the textile-specific parameter
- buffering capacity against vaporous sweat impulses F_d (moisture regulation index), which should be high.

(3) During *heavy sweating situations*, a high amount of liquid sweat appears on the skin. As liquid sweat is transported by other physical processes (capillary effect, adsorption and migration) than water vapour [15, 17], different Skin Model tests are required [16]. They measure
- the buffering capacity against liquid sweat impulses K_f which ranges between 0 and 1 and is defined as the ratio of the amount of sweat which is taken away from the skin over the amount which was originally present on the skin;
- the liquid sweat transport F_l, which should be high.

(4) Last but not least, the wear situation directly *after an exercise* is also of great relevance to e.g. sport textiles. Then, the textile might be soaked up with sweat and has lost its thermal insulation. This leads to the so-called postexercise chill, which is very unpleasant. The postexercise chill is characterized according to BPI 1.3 [18] by the parameters:
- water retention ΔG, which should be small;
- thermal insulation of wetted textile R^*_{ct}, which should be high;
- drying time Δt, which should be short.

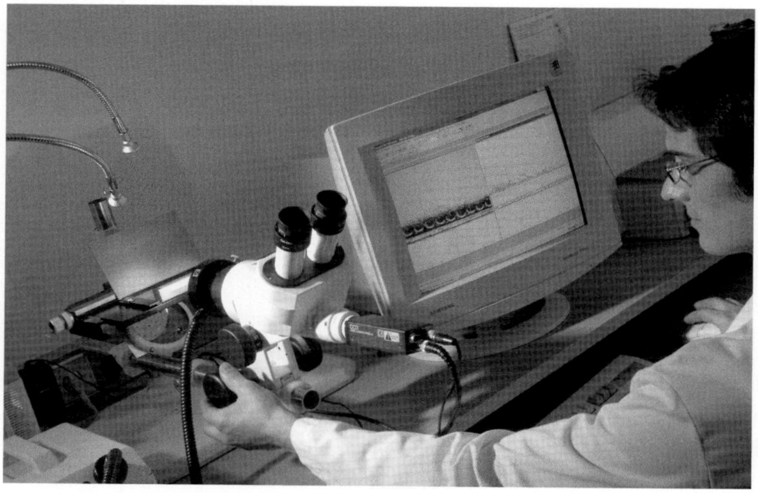

Fig. 2. Measurement of the surface index i_O.

Skin Sensorial Test Apparatus

Also the skin sensorial wear comfort of textiles can be tested by a set of special laboratory apparatus [19, 20]. Within this work, the following textile-specific parameters have been measured: (1) the wet cling index i_K, (2) the sorption index i_B, (3) the surface index i_O, (4) the number of contact points with the skin n_K and (5) the fabric's stiffness s.

As an example, in figure 2 the measurement of the surface index i_O is shown.

Wear Comfort Vote

From the Skin Model measurements as well as from the skin sensorial tests, thermophysiological and skin sensorial wear comfort votes can be calculated, respectively [21, 22]. They range from 1 = 'very good' to 6 = 'unsatisfactory'. The thermophysiological and the skin sensorial wear comfort votes can be combined to an overall wear comfort vote, predicting the perceived wear comfort in practice. Differences of 0.5 or more can be regarded as perceivable.

The thermophysiological wear comfort vote for woven fabrics used e.g. in occupational clothing (work wear) or trousers is calculated by

$$WC_T = 0.576\,R_{et} - 2.314\,F_d - 3.1613\,K_f - 0.0823\,F_l + 0.0341\,\Delta G + 4.83$$

The one for 'normal' underwear worn in everyday situations is given by

$$WC_T = -5.64\, i_{mt} - 0.098\, F_i - 2.248\, F_d - 4.532\, K_f + 10.8$$

The skin sensorial wear comfort votes are

$$WC_S = -2.537\, i_{mt} + 0.0188\, i_K + 0.00229\, i_B + 0.0209\, i_O + 0.00171\, n_K + 0.0386\, s - 1.64 \text{ for woven fabrics and}$$

$$WC_S = -2.537\, i_{mt} + 0.0188\, i_K + 0.00229\, i_B + 0.0209\, |9 - i_O| + 0.00171\, n_K + 0.0386\, |16 - s| + 0.36 \text{ for knitwear.}$$

For woven fabrics intended as work wear as well as for underwear the overall wear comfort is derived by the formula

$$WC = 0.66\, WC_T + 0.34\, WC_S$$

For sports underwear, which is frequently in contact with liquid sweat, the overall wear comfort is

$$WC = -0.171\, F_l + 0.293\, \Delta t - 16.047\, i_{mt} - 0.153\, F_i + 0.449\, WC_S + 2.649$$

Wearer Trials

Within the research project wearer trials with human test subjects were performed under controlled climatic and activity scenarios in a climatic chamber. Via probes attached to the subject's body, various objective data can be obtained. The data presented here comprise heart rate, rectal temperature, skin temperatures at different positions and humidity in the microclimate. An example for these types of test is given in figure 3.

Materials

The textiles investigated are given in table 1. They differ in their application (work wear, trousers, underwear) and their biofunctional treatment (Cu_2S, silvered copper monofilaments, cyclodextrin finishing, different fibres with bacteriostatic agent in the polymer matrix).

As a reference, for each biofunctional treatment an identically constructed but not biofunctional textile was investigated, too. It should be mentioned that, as the wear comfort is not only affected by the biofunctionality of the textiles, but also by their different constructions, one can only deduce the influence of

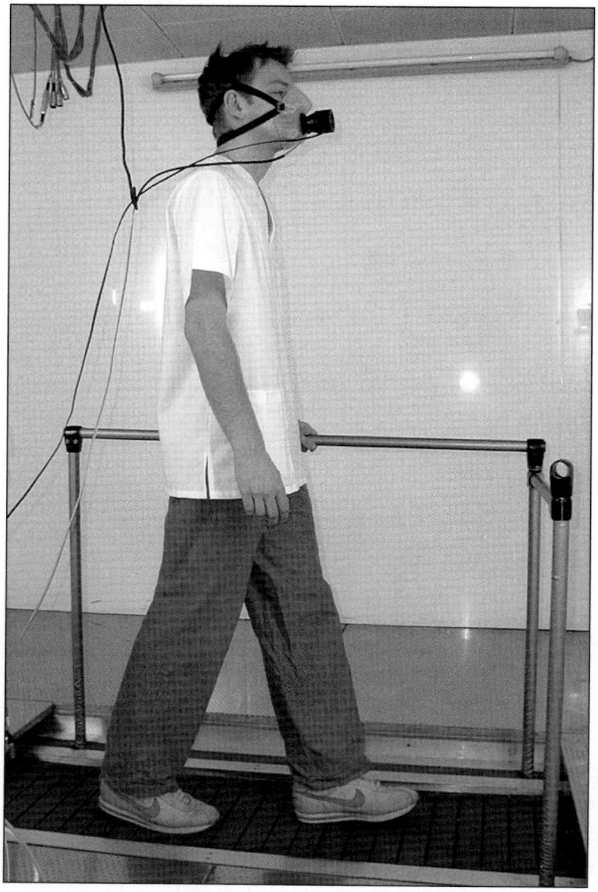

Fig. 3. Photo of a test person during wearer trials with tunics. The face mask is needed to determine the metabolic rate by analysing the exhaled air's oxygen and carbon dioxide contents.

the biofunctionality in comparison with its reference material, but not with another biofunctional textile which is differently constructed.

Results and Discussion

Material Tests

It soon turned out that, from a physiological point of view, different biofunctional modifications are not comparable. Some cause a clear and perceivable worsening of the wear comfort, whereas others have no influence at all.

Table 1. Description of the investigated textile samples

No.	Application	Material
1	work wear	as No. 4 but weft with PA 6.6 threads covered by Cu_2S
2	work wear	as No. 4 but with silvered copper monofilaments in a 2-mm grid
3	work wear	as No. 4 but with cyclodextrin finishing
4	work wear	twill weave 2/1, PES/CO 65/35, 215 g/m^2 (reference)
5	trousers	as No. 6 but with cyclodextrin finishing
6	trousers	twill weave 3/1, CO/EL 98/2; warp: CO; weft: CO/EL (reference)
7	underwear	PA fibres with bacteriostatic agent in polymer matrix (fibre inherent) based on silver
8	underwear	as No. 7 but with conventional PA (reference)
9	underwear	PES/PAN, PAN fibre with bacteriostatic agent
10	underwear	PES, as No. 9 but without biofunctional PAN (reference)
11	underwear	single jersey, Nm70 CO/CMD 67/33, 150 g/m^2, antibacterial substance incorporated into CMD fibres
12	underwear	as No. 11 but with non-biofunctional CMD (reference)
13	underwear	PES/CO 65/35, PES fibres with Ag ions on a ceramic substrate
14	underwear	As No. 13 but with non-biofunctional PES (reference)
15	work wear	PES/CO 65/35, PES fibres with Ag-ions on a ceramic substrate
16	work wear	as No. 15 but with non-biofunctional PES (reference)

CMD = Modal; CO = cotton; EL = elasthane; PA = polyamide; PAN = acrylic; PES = polyester.

Interestingly, not only the finishes, but also the fibre-inherent biofunctionality may cause clear differences to normal, non-biofunctional textiles.

In table 2, the skin sensorial and the overall wear comfort votes are given. In each case the biofunctional textile is compared to its reference material. It can be seen that especially the skin sensorial comfort may be significantly worsened by the biofunctionality of the textile. Starting with sportswear, sample No. 7 with a bacteriostatic agent has to be judged as 'poor' concerning its skin sensorial properties, whereas the reference material is at least 'satisfactory'. Also No. 9 worsens the skin sensorial wear comfort vote. This specimen is only rated as 'sufficient', but the reference material is 'satisfactory'. These differences can be clearly perceived by a wearer. On the other hand, the modified polyester fibre with inherent silver ions on a ceramic substrate does not lead to a significant worsening of the skin sensorial wear comfort (samples 13 and 14).

The modified polyester fibre does not deteriorate the overall wear comfort either. The textile construction of sample No. 9 is good enough to mask the worsening caused by the biofunctional acrylic fibre so that it gets the same very good rating as its reference material. But sample 7 only offers a 'satisfactory' overall wear comfort, whereas its reference is to be judged as 'good'.

Table 2. Skin sensorial and overall wear comfort votes WC_s and WC of biofunctional textiles and reference materials

Sample No.	WC_s	WC
1	3.5	2.3
2	2.7	2.3
3	3.1	2.2
4	2.5	2.1
5	2.8	3.1
6	2.2	2.8
7	4.8	3.0
8	3.3	2.0
9	3.9	1.0
10	2.7	1.0
11	2.6	2.1
12	2.4	2.0
13	1.9	1.5
14	1.6	1.6
15	2.2	1.9
16	2.3	1.9

The accuracy of WC_s and WC is 0.3, differences ≥ 0.5 can be subjectively perceived by a wearer.

It is interesting to discuss now why some of the biofunctional modifications worsen the wear comfort. In the case of the sportswear samples No. 7 and 9, the reason is a hydrophobic fibre surface. Textile No. 7 needs more than 9 min to soak up a single sweat drop, No. 9 does not take it up at all during a 10-min test. But their reference materials 8 and 10 just need $i_B = 25$ or 12 s, respectively, to take up the same amount of sweat.

The hydrophobic biofunctional textiles cause problems particularly in the liquid sweat transport and lead to a moist skin, which can be easily irritated. In addition, the hydrophobic surface is perceived as clinging more to sweat-wetted skin (higher wet cling index i_K). As clinging feels very ugly, the wearer subjectively recognizes the difference to the hydrophilic reference materials.

Discussing the results for normal underwear, which is not intended to be used during sports but in everyday situations (samples 11 and 12), one can see that the antibacterial substance incorporated into the Modal fibres does not affect the wear comfort. Within the accuracy of the prediction, both materials have to be judged as equivalent regarding their skin sensorial and overall wear comfort vote.

In the case of work wear, the inclusion of biofunctional PA 6.6 filament yarns covered by Cu_2S, as used in sample 1, was found to be disadvantageous for the skin sensorial comfort (compared to the reference material 4). Again, the biofunctional material gets hydrophobic, probably due to the Cu_2S. In addition, the use of non-textured filament yarns instead of spun yarns leads to a smoother (flatter) textile surface, which has too many contact points with the skin n_K and clings more easily to sweat-wetted skin. However, there is only little influence on the overall wear comfort.

In principle, the silvered copper monofilaments as used in sample 2 also lead to a smoother surface and more contact points. But as these yarns are used only in a 2-mm grid, their number is too small to have a negative effect. In fact, the wettability index i_B and the wet cling index i_K are even slightly better than those of the reference material 4. Hence, this grid construction does neither worsen the skin sensorial nor the overall wear comfort.

If cyclodextrins are applied to textiles, the exact finishing process is crucial for the wear comfort. Some of the finishes lead to cross-linking between the cyclodextrin molecules, causing a stiffening of the textile, as it is observed for samples 3 (work wear) and 5 (trousers), which is again problematic from a skin sensorial point of view. Additionally, these cross-linking cyclodextrin finishes effect a more hydrophobic textile surface. Consequently, their skin sensorial wear comfort vote is worsened. Although the overall wear comfort vote is just affected too little to be perceivable, especially the greater stiffness may be problematic.

These problems with cyclodextrin textiles can be overcome by special finishes, which selectively link the cyclodextrin molecules to the fibres [2]. However, also the finishing recipe is important, because overdosing would again lead to cross-linking between cyclodextrins.

Last but not least, as for use in sportswear, the biofunctional polyester fibres with Ag ions on a ceramic substrate do not negatively influence the wear comfort of work wear (samples 15 and 16).

Wearer Trials with Test Persons

To verify that biofunctional textiles can indeed offer the same wear comfort as conventional, non-biofunctional ones, wearer trials with human test subjects have been performed. As an example, antimicrobial work wear for hospital staff was chosen. It is intended to have a better protection of the wearer against germs and to reduce germ spread. Whether biofunctional textiles really offer these hygienic advantages or not cannot be decided here but has to be discussed by hygienists. But in any case, antimicrobial clothing is only accepted by hospital staff, if it offers a sufficient wear comfort.

For the wearer trials presented here, tunics were tailored from the materials 15 (biofunctional) and 16 (reference). Four test subjects wore each sample twice (1 repetition, 8 trials per specimen). The test persons performed activity/rest cycles consisting of walking on the treadmill with 5 km/h and sitting,

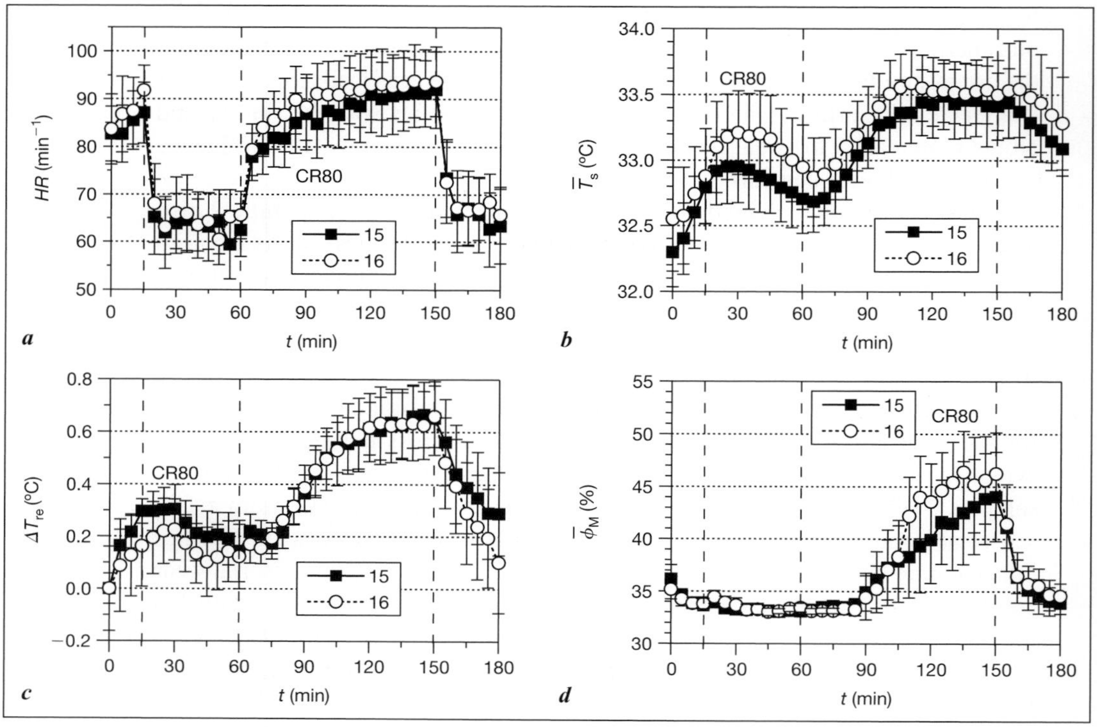

Fig. 4. Time-dependent objective data from trials with human test subjects wearing tunics; No. 15 = biofunctional, No. 16 = reference; CR80 = 80% confidence range. ***a*** Heart rate *HR*. ***b*** Mean skin temperature $\overline{T}_s$ ***c*** Increase in rectal temperature ΔT_{re}. ***d*** Mean relative humidity in the microclimate $\overline{\phi}_M$.

simulating alternating work conditions in a hospital. Ambient temperature and relative humidity were $T_a = 23°C$ and $\phi_a = 50\%$, respectively. A photo of the tests is shown in figure 3.

In figure 4, objective data of the increase in the rectal temperature ΔT_{re}, the heart rate *HR*, the mean relative humidity in the microclimate $\overline{\phi}_M$ (averaged over all body positions) and the mean skin temperature $\overline{T}_s$ are given as functions of time *t*. It can be seen that all curves for both samples are lying closely together. Consequently, there are only a few significant differences, which are too small to be perceived by the wearer at any time.

These objective data are confirmed by the subjective sensations of the test persons given in figure 5. It is displayed that both tunics are judged approximately with mark 2 ('good'). Thus, it has to be concluded that the antimicrobial textile offers the same good wear comfort as the conventional one.

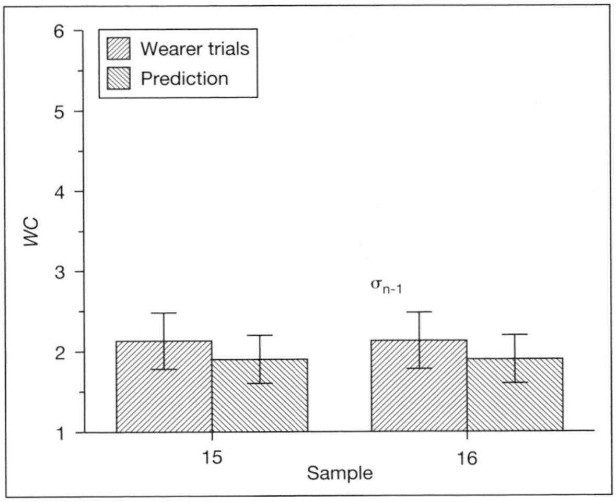

Fig. 5. Wear comfort vote *WC* as judged by human test subjects and as predicted from Skin Model and skin sensorial tests according to the section 'Wear comfort vote'.

In figure 5, the comparison with the predicted wear comfort vote based on the Skin Model and the skin sensorial measurements is presented as well. Obviously, the predictions are correct: Not only the relative assessment that both materials offer the same wear comfort, but in particular the absolute values nicely coincide with the statements of the test persons. Within the accuracy of the measurements and the statistical distribution of the group of test persons (having slightly different opinions) both data sets are equal (mark 2 'good').

Information of the Customer

Producers who are able to convince the end user of their product's benefits like comfort or a biofunctional function, in particular directly at the point of sale, have an advantage. Therefore, most apparels come with hang tags claiming extraordinary wear comfort or functions. However, in the shop, the consumer is not able to distinguish between real benefits and sole marketing gimmicks. But a consumer who has been disappointed once is difficult to convince to buy expensive high-tech clothing again.

Thus, from the consumer's and the high-quality producer's point of view, there is a need for an independent testing of the apparel's physiological and biofunctional properties. At least in the field of protective clothing, many of these

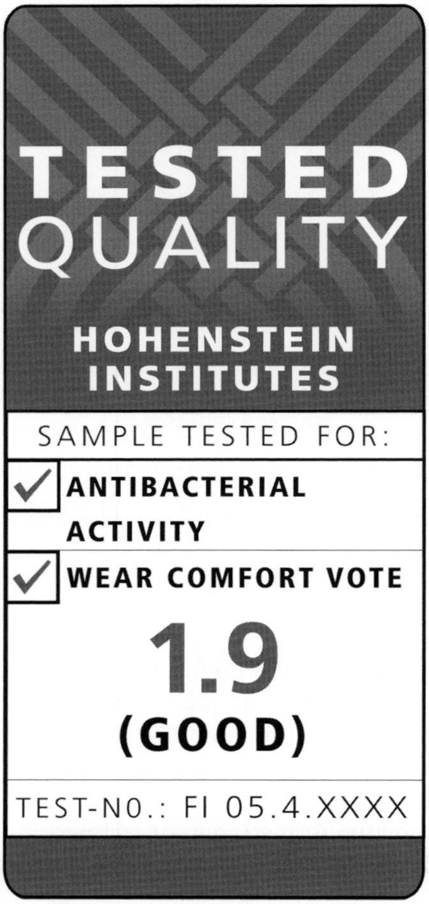

Fig. 6. The Hohenstein Institutes' label 'Tested Quality' as an example of how to display product benefits like wear comfort and antibacterial activity.

garments have to be certified by EU-notified bodies, in order to guarantee the end user a certain level of protection and comfort.

However, for many applications of biofunctional textiles like sportswear, there is no European standardization in sight. But still producers of high-quality textiles and clothing may wish to show that their product is independently tested, e.g. for wear comfort or a biological function. An option is shown in figure 6, the Hohenstein Institutes' label 'Tested Quality'. By this, apart from other properties like wind or water tightness, in particular physiological or biofunctional properties of the clothing product can be advertised. This label may also be used by textile producers to show the benefits of their materials to their direct customers like garment producers.

Conclusions

Today, the physiological wear comfort is of great importance, and biofunctional textiles only have a high market potential, if they are comfortable. It was shown how the wear comfort can be measured effectively by a set of clothing physiological laboratory test apparatus. The resultant, predicted wear comfort votes match exactly the subjective perceptions of test persons. As a result, biofunctional textiles can offer the same good wear comfort as classical, non-biofunctional materials. On the other hand, some of the biofunctional treatments lead to a perceivably poorer wear comfort. Guidelines for the improvement of the thermophysiological or skin sensorial wear comfort were given.

Acknowledgement

We are grateful to the Forschungskuratorium Textil for the financial support of the research project (AiF No. 12852), which was funded by the German Ministry of Economy and Work via grants of the Arbeitsgemeinschaft industrieller Forschungsvereinigungen Otto von Guericke.

References

1 Mucha H, Höfer D, Swerev M: Antimikrobielle Bekleidungstextilien im Trend – Funktion, Einsatzgebiete und medizinische Bewertung; in Groth UM, Kemper B (eds): Jahrbuch für die Bekleidungswirtschaft 2003. Berlin, Schiele & Schön, 2003, pp 29–37.
2 Buschmann HJ, Denter U, Knittel D, Schollmeyer E: The use of cyclodextrins in textile processes – An overview. J Textile Inst 1998;89:554–561.
3 Elsner P, Kurz J (eds): European Conference on Textiles and the Skin. Apolda, April 11–13, 2002.
4 Bartels VT, Umbach KH: Physiological function and wear comfort of protective clothing with the examples of OR gowns and fire fighter garments. 4th Int Conf Safety Protect Fabrics, Pittsburgh, October 26–27, 2004.
5 Foster L: Sportswear 2000 – Interpreting fabric trends. World Sports Activewear 1998;4(3):21–24.
6 Ullsperger A: Innovation strategy of smart textiles products and high-tech fashion. Textile Int Forum Exhibition, Taipei, 2001.
7 Reinhold K: Wie viel Funktion darf's denn sein? Textilwirtschaft 2001;56(48):70–72.
8 Albaum M: Der Stellenwert der Funktion im Bekleidungshandel; in Knecht P (ed): Funktionstextilien. Frankfurt, Deutscher Fachverlag, 2003, pp 83–94.
9 Bartels VT: Grundsatzuntersuchung der bekleidungsphysiologischen Eigenschaften biofunktioneller Textilien. Tech Rep No AiF 12852. Bönnigheim, Hohenstein Institutes, 2003.
10 Mecheels J: Körper – Klima – Kleidung: Wie funktioniert unsere Kleidung? Berlin, Schiele & Schön, 1998.

11 Mecheels J, Umbach KH: The psychrometric range of clothing systems; in Hollies NRS, Goldman RF (eds): Clothing Comfort. Michigan, Ann Arbor Science, 1977, pp 133–151.
12 Bartels VT, Umbach KH: Messverfahren zur Beurteilung der Atmungsaktivität von Textilien für Bekleidung und Bettsysteme. Melliand Textilber 2003;84:208–210.
13 ISO 11092, EN 31092: Measurement of thermal and water-vapour resistance under steady-state conditions (sweating guarded-hotplate test). Geneva, International Standards Organization, 1993.
14 Umbach KH: Measurement and evaluation of the physiological function of textiles and garments. 1st Joint Conf 'Visions of the Textile and Fashion Industry', Seoul, 2002.
15 Umbach KH: Moisture transport and wear comfort in microfibre fabrics. Melliand English 1993;74:E78–E80.
16 BPI 1.2: Measurement of the buffering capacity of textiles with the thermoregulatory model of human skin (Skin Model). Bönnigheim, Hohenstein Institutes, 1994.
17 Umbach KH: Optimization of the wear comfort by suitable fibre yarn and textile construction. 40th Int Man-Made-Fibres Congr, Dornbirn, 2001.
18 BPI 1.3: Measurement of thermal insulation of a wetted fabric with the thermoregulatory model of human skin (Skin Model). Bönnigheim, Hohenstein Institutes, 1985.
19 Bartels VT, Umbach KH: Skin sensorial wear comfort of sportswear. 40th Int Man-Made-Fibres Congr, Dornbirn, 2001.
20 Bartels VT, Umbach KH: Test and evaluation methods for the sensorial comfort of textiles. Euroforum 'Toucher du Textile', Paris, 2002.
21 Umbach KH: Bekleidungsphysiologische Gesichtspunkte zur Entwicklung von Sportkleidung. Wirk Strick Tech 1993, vol 43(2).
22 Bartels VT, Umbach KH: Sehr gut: Noten für den Tragekomfort. Kettenwirkpraxis 2003;37(1):30–33.

Volkmar T. Bartels
Hohenstein Institutes, Department of Clothing Physiology
DE–74357 Boennigheim (Germany)
Tel. +49 7143 271 611, Fax +49 7143 271 94611, E-Mail v.bartels@hohenstein.de

Burg G (ed): Biofunctional Textiles and the Skin.
Curr Probl Dermatol. Basel, Karger, 2006, vol 33, pp 67–77

Antimicrobial Textiles, Skin-Borne Flora and Odour

Dirk Höfer

Institute for Hygiene and Biotechnology, Hohenstein Research Center, Boennigheim, Germany

Abstract

Along with climate and physical activity, textiles have an effect on sweating and the development of odours. Accordingly, textiles inadequately optimized in terms of clothing technology as a result of poorly cut structures or poor materials result in increased sweating and odour. However, the development of body odour itself cannot be avoided, even with optimally designed clothing. Therefore new textiles, 'treated with antimicrobial agents', have been developed, with the aim of reducing odour by decreasing the number of germs on the skin. From the scientific point of view, the interactions between textiles, sweat, skin and skin flora are extremely complex. For this reason, this article explains in more detail the basic principles of odour formation resulting from sweat and how this can be influenced by textiles treated with antimicrobial agents. With reference to the results of recent research, the article looks into questions of how textiles treated with antimicrobial agents have an effect on populations of skin bacteria.

The Physiology of Sweating and the Development of Odour

For odour to originate from sweat, it is essential that certain germs on the skin be involved. This can be seen already by the changing odour profile obtained after the application of substances with no odour but with an antibacterial effect, such as deodorants. The majority of sweat is formed in what are termed the sweat gland coils in the dermis (eccrine sweat glands, fig. 1), approximately 2–3 million of which are distributed over our entire body [1]. This eccrine sweat is a clear, watery and almost odourless liquid made up of 99% water, which evaporates on the skin and cools the body in the process. This means that eccrine sweat has a direct part to play in the measurable wear comfort of textiles in terms of clothing

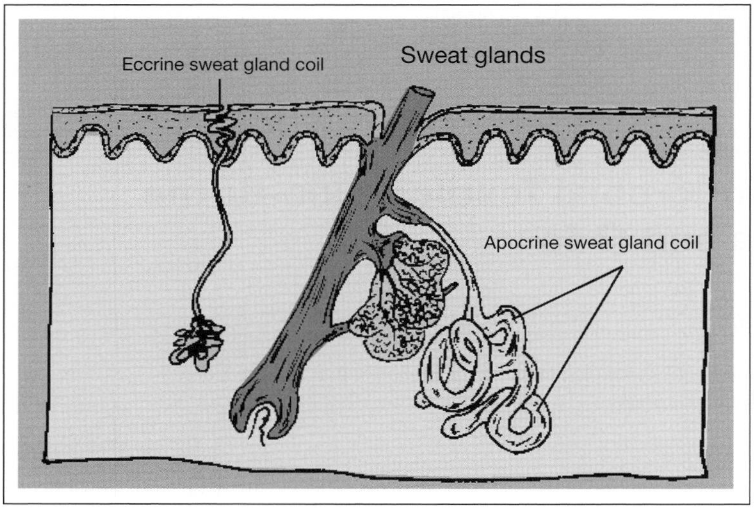

Fig. 1. Scheme of human sweat glands.

physiology ('breathing activity'). However, the purpose of this article is rather to explain the development of the pungent, unpleasant sweat odour, an odour profile arising out of another type of sweat gland, the odour glands also known as apocrine sweat glands (fig. 1). These occur alongside the eccrine glands in various parts of the body, the armpit hair follicles, eyelids, the aural and nasal passages, around the nipples and in the pubic and anal areas. The duct aperture of the odour glands is not at the skin surface but at the base of the hair follicle. Odour glands are most frequently found in the human armpit at a ratio of 1:1 with eccrine sweat glands. However, even the freshly produced apocrine sweat is initially completely odourless when it reaches the surface of the skin. Only when exposed to the effects of certain bacteria of the skin flora does the characteristic, more or less unpleasant body odour arise.

The purpose of textiles treated with an antimicrobial agent is to achieve a reduction in body odour by restricting the growth of these bacteria, a task which clothing worn next to the skin is, of course, better able to fulfil than outer clothing, as it is in closer contact with the germs on the surface of the skin.

The Skin: Protective Organ and Environment for Skin Germs

To gain a better understanding of the way antimicrobial textiles work, it is worth taking a close look at the environment occupied by our skin flora, the

Table 1. The human skin flora

Occurence	Type	Bacteria	Dominant species
Moist and dry areas of skin	staphylococci micrococci	aerobic gram-positive cocci	staphylococcus epidermidis et hominis
Moist and dry areas of skin	coryne bacteria brevi bacteria	aerobe coryneforme rods	
Hair follicles with many sebaceous glands	proprioni bacteria	anaerobic coryneforme rods	proprionibacterium acnes
Hair follicles with many sebaceous glands	malassezia furfur	yeast fungi	pityrosporum ovale

distribution of odour-causing micro-organisms and how they work. Skin flora micro-organisms have evolved by adapting to their living conditions on and in the skin [2]. Accordingly, the relevant bacterial flora on the skin is generally relatively stable. Typical (locally resident) skin flora includes aerobic micro-organisms such as staphylococci, coryneform bacteria, micrococci, yeasts and propionibacteria (table 1). To date, there is no indication that illness-causing (pathogenic) germs play a part in the creation of odour. Nevertheless, pathogenic germs are kept away from the environment of the resident germs by a variety of defence mechanisms: in the first instance, the horny layer, the outer layer of the epidermis, represents an effective protective system, and at the same time a significant mechanical barrier against pathogenic micro-organisms. It also provides the skin with the greatest degree of protection against a variety of chemical and physical influences from the surrounding environment. The horny layer is made up of some 15–20 layers of cells in the form of dead skin cells, termed horn cells; these are embedded in various fat molecules in the outer skin (cholesterol, free fatty acids, ceramides), along with which they form an effective permeability barrier. The skin's protective system against pathogenic bacteria and fungi also includes the hydrolipid film, an emulsion of water and fat covering the horny layer, keeping the surface of the skin supple and the skin itself moist. The hydrolipid film is made up of sweat and sebaceous-gland fats, along with substances from the horn-producing process, e.g. horn cells. Due to the proportion of slightly acidic components such as lactic acid, pyrrolidone carbon and amino acids, the hydrophilic proportion of the hydrolipid film represents the so-called acidic mantle of the skin, displaying physiological pH values of between 5.4 and 5.9. Staphylococci, coryneform bacteria and micrococci prefer to remain in the hydrolipid film; however, they are also found more deeply in the skin, in the top layer of the horny layer cells. Yeasts and anaerobic coryneform bacteria of the species *Propionibacterium* occur predominantly in the sebaceous-gland ducts and hairs (table 1). The micro-organisms of the resident skin flora support the low pH value of the acidic mantle through the additional production of low-molecular-weight fatty acids, by breaking down organic components of the sweat and the hydrolipid layer. These bacterial metabolic products cause the typical, slightly acidic odour of the actual skin, but not the pungent, unpleasant smell of sweat. At the same time, the acid pH value of the acid mantle wards off Gram-negative bacteria (e.g. Enterobacteriaceae, primarily gut germs) from the surface of the skin.

As an additional protective system for the skin, eccrine sweat contains the body's own antimicrobial substances, such as immunoglobulins, lysozyme (an enzyme which attacks the bacterial wall of the germ) and dermicidin, a 47-amino-acid-long peptide exerting an antimicrobial influence on a large number of pathogenic micro-organisms [3]. Under the horny layer, the living skin cells (keratinocytes) also produce antimicrobial substances (peptides such as cathelicidin, secretory leukoprotease inhibitor, the C-terminal fragment of cathelicidin LL-37, psoriasin, β-defensin) [3, 4], especially when the skin is defending itself against pathogenic germs which have made their way in (e.g. in the case of psoriasis) [4]. Finally, pathogenic germs are also kept away from the surface of the skin by special antimicrobial peptides (bacteriocins) emitted for the defence of the micro-organisms of the resident skin flora [5]. Accordingly, in the above competition, the resident germ population can, for the most part, successfully hold its own over foreign germs, such as *Staphylococcus aureus*, species from the bacillus family, pseudomonads and Enterobacteriaceae.

Odour Formation and Odour Molecules

Human skin flora displays clear intra- and interindividual differences and, accordingly, different odours, i.e. the typical armpit odour which is different from the slightly acidic skin odour of eccrine sweat. In the moist, warm armpit in particular, micro-organisms find ideal living conditions. At the same time, the type of bacteria is the determining factor for human odour. In bacteriological terms, at present a distinction is made between 2 types of armpit germ flora: those in which coryneform bacteria dominate and those dominated by micrococci. Micrococci tend to give rise to a weaker 'acidic' body odour, an odour type occurring particularly frequently among women. Conversely, the 'pungent, biting' type of odour occurs to a greater degree among men. It is caused by the effects of lipophilic corynebacteria which occur predominantly (also in terms of quantities) among males. At the same time, the intensity of the smell is basically determined by the number of bacteria. The average density of bacteria in the case of coryneform flora is 6 times higher than under coccal domination, meaning that women generally generate less armpit odour than men. Apart from personal hygiene, of course, the food available (nutrition), varying levels of sebaceous-gland production and differing degrees of moisture also play an important part in the composition of personal skin flora. If, e.g. what are normally dry areas of skin are kept within a moist 'subtropical' climate (e.g. by textiles displaying little breathability), then an explosive increase in the bacterial density of coagulase-negative staphylococci and coryneform bacteria up to about 5 factors of 10 can be established [6].

Coryneform rod bacteria convert decomposition products of the male sex hormone testosterone from armpit sweat [7]. Women also produce quantities of testosterone in sweat, even if to a lesser degree [8]. Initially, the bacteria convert the testosterone derivative androstadienone – eliminated along with the sweat – to form androstenol so that, in a subsequent step, it can be further processed into androstenone, a molecule to which the female nose seems to be quite specifically sensitive. From the group of the many volatile C6–C11 long fatty acid molecules, to date it has only been possible to identify isovalerian acid and 3-methyl-2-hexenoic acid as significant odour molecules, in terms of quantity. Short-chain fatty acids and isovalerian acids are also the main odour-creating components of foot sweat [9, 10]. Interestingly, the fatty acids from the depths of the apocrine sweat glands initially reach the outer surface of the skin by binding onto special carrier proteins and are only released at the surface by the bacterial effects of these proteins, to emerge as odoriferous material [11].

The aversion to the smell of sweat is a sociological and cultural phenomenon. Who likes to be in the company of a 'strongly smelling' companion? Throughout evolution, individual body odour has played a significant role in social behaviour. So what makes the body odour of two people so individual? Apart from different skin flora and the subsequent production of fatty acids, this is presumably due to the part played by proteins in the specific body scent. Accordingly, it is known that the transplantation proteins specific to each person (MHC proteins) play a part in the creation of an individual scent signature [12]. MHC transplantation proteins are found as individual 'identifying proteins' on the surface of certain immune cells (MHC II) and other body cells (MHC I). It is largely unknown how these proteins help to influence body odour. Apart from MHC proteins, there is often discussion of hormone-type scent materials, so-called pheromones, as possible odour molecules [13]. While these scent molecules have an interesting role to play in the animal kingdom in terms of sexual and social behaviour, it is presumed that they do not play a part in the personal body scent of the human being, as, to date, there is no unambiguous scientific evidence that pheromones have any effect at all on humans. In addition, in the animal kingdom pheromones are not consciously identified by the nasal mucous membrane – and therefore as a scent signature – so if nevertheless they should occur in humans, in any case pheromones may be registered subconsciously.

Safety Aspects of the Interaction between Textiles and Skin Flora

In the past, there have been questions raised by dermatologists and hygienists in terms of safety in the use of textiles treated with antimicrobial agents. Given the complexity of the interactions between textiles and the skin along with our skin flora (see above), these questions are, of course, not easy to answer. What can be said about the chemical-toxic effects of the biocide agent used on our skin or on the human organism is that these influences are, of course, dependent on the specific human toxicological properties of the relevant substance used as the agent. With the use of preparations with a silver content, most textile processors have preferred a treatment surrounding where there are no reservations in terms of human toxicity, with this even being successfully put to therapeutic use on the sensitive atopic dermatitis skin [14, 15]. The earlier use of tri-butyl tin has, in the meantime, largely disappeared from the market. A further question is directed at possible secondary risks due to the influence exerted on the flora balance. In this connection, a few scientific comparisons have to be discussed.

How Intensive Is the Contact between Antimicrobial Textiles and the Skin Flora?

Trials involving wearing textiles treated with antimicrobial agents to avoid body odour show that reducing the Gram-positive germ population of the resident skin flora, especially in the above-mentioned problem areas (armpits, feet), results in a reduction of body odour. To record the reduction in germs brought about by antimicrobial textiles, test systems containing Gram-positive test germs are therefore appropriate, e.g. the JIS 1902-2002 quantitative suspension test [16]. In this test, the test germs are introduced into a reaction vessel for 18 h, in permanent and direct, intensive contact with the test sample so that almost all of the fibres are in close contact with the germs. This allows the maximum achievable 'efficiency rating' of the treated textiles to be quantitatively recorded as 'growth inhibition'. In its structure, the test is not designed as a 'textile-on-skin wearing test', as even textiles worn next to the skin only make contact with the skin at some sites, and not over the full surface, forming individual points of contact instead [17]. The degree to which antimicrobial-treated textiles make contact with the skin, and therefore with the skin flora, depends on the relevant textile construction: depending on the fibre or thread titre, thread structure or type of surface profile there is, for example, a different number of contact points $n_{K1/C1}$ formed with the skin [17]. The adhesion to skin soaked with sweat, which can be recorded as adhesive force index $i_{K1/C1}$, has an influence as a textile parameter on the contact between treated textiles and the skin. Therefore, in the final analysis, the question can only be answered by long-term wearing trials in the course of which a variety of textile constructions involving textiles treated with antimicrobial agents are used.

What Efficiency Rating Is Achieved by Textiles Treated with Antimicrobial Agents on Skin Flora?

In the JIS 1902-2002 suspension test, after 18 h of incubation some textiles achieve a reduction in growth in the germ suspension of around 3 powers of 10. What significance should be attributed to this figure? If we take into account that, by definition, disinfectants must achieve germ reductions of >5 log levels, it can be deduced that antimicrobial textiles cannot operate as 'skin disinfectants'. This would be an incorrect assumption. Rather, these textiles reduce the germ population to a certain degree. If one compares the textile situation with clinical studies in which, e.g. an area of the skin is disinfected with disinfectant before an operation, it can be seen that the germ population of the skin is only reduced in the short term but by no means disappears [2]. After a certain time, new skin micro-organisms germinate out of the deeper skin deposits mentioned above (sweat pores, hairs, outer skin layers) and make up the 'deficit' in the skin flora on the surface. So from the scientific point of view, the same effect should be anticipated from wearing antimicrobial textiles.

What Effective Range Is Achieved by Textiles Treated with an Antimicrobial Agent?

Basic wear tests have been conducted to look for the effective range of bacteria found in antimicrobial textiles. In the first instance, the question of the extent to which textiles can be colonized by typical skin germs on the emission of eccrine sweat was addressed. Accordingly, test subjects wore a synthetic polyamide (PA) material in a 6-hour wearing trial under non-occlusive conditions on the side of the chest. The textile was then incubated in DAPI solution in order to mark the DNA of the adhering skin germs. Under the fluorescence microscope in figure 2 a colony of typical skin germs (*Staphylococcus epidermidis*) can be seen, adhering between two parallel PA fibres. With reference to the sample surface of 1.8 × 1.8 cm used, the colony in the test fibres must be described overall as very small. However, given the period of time of 6 h and the low level of nutrition available (eccrine sweat is relatively low in nutritional value), this is hardly surprising. There was preferred colonization of parts of the textile especially in the area of adhering skin flakes (greater amount of nutrient available compared

Fig. 2. Fluorescence image of two parallel PA fibres with DNA-labelled skin germs.

with PA) if the textile was incubated for a further 18 h at 36°C after the wearing test. Here, there was a clear increase in bacteria observed, especially in the area of skin flakes adhering to fibres (not shown).

As the nutrient available on/in the textile is clearly a crucial parameter for the growth and colonization of certain germs, non-treated textile fibres on an agar plate with a nutrient content were inoculated with germs capable of forming biofilms (*Klebsiella pneumoniae*). In these studies, there was a change in the adhesive behaviour of the germs. In a fluorescent image, figure 3 shows a clear biofilm structure created by *Klebsiella* germs along a non-treated textile fibre.

The question as to the effective range of treated textiles – and, consequently, the influence exerted on skin flora – is posed both in respect of odour development (interaction between body bacteria, sweat and the textile) as well as the possible use of textiles treated with an antimicrobial agent for the prevention of infection (interaction between airborne germs, or, as the case may be, foodstuffs, organic components and the textile). In order to identify the initial reference points here, PA textile fibres treated with antimicrobial agent (e.g. in the form of silver impregnation) were placed into a bacterial colony of *K. pneumoniae* and then marked with Syto 9 fluorescent dye which penetrates through the cell membrane of living germs and stains the inside of the bacterium. Under the fluorescence microscope, this resulted in the focus level in a large number of fluorescence-marked bacteria rods, for the most part occupying a position outside an inhibition zone alongside the textile fibre. Morphometric measurement of the fibre showed a fibre average of 100 μm, meaning that in this case it is possible to deduce a 50- to 100-μm-wide inhibition zone along the fibre (fig. 4a). At the

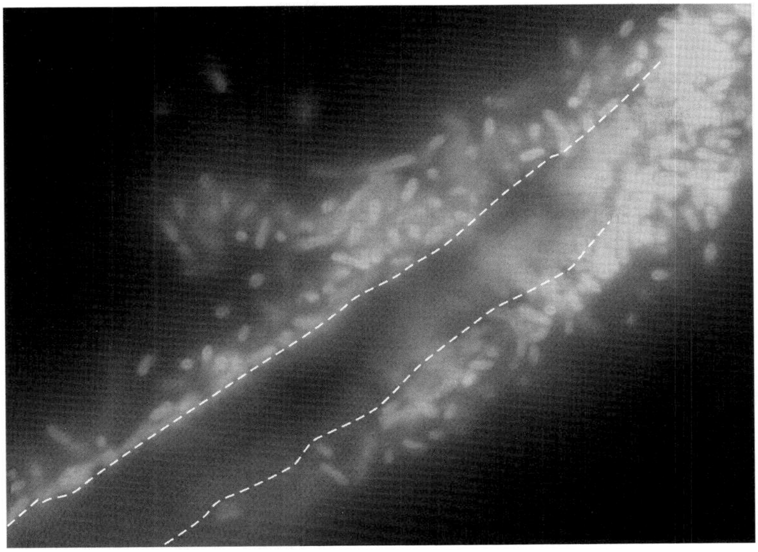

Fig. 3. Biofilm-producing micro-organisms, attached to a man-made fibre, labelled with the DNA marker Sytox green. Dashed line: PA fibre.

a 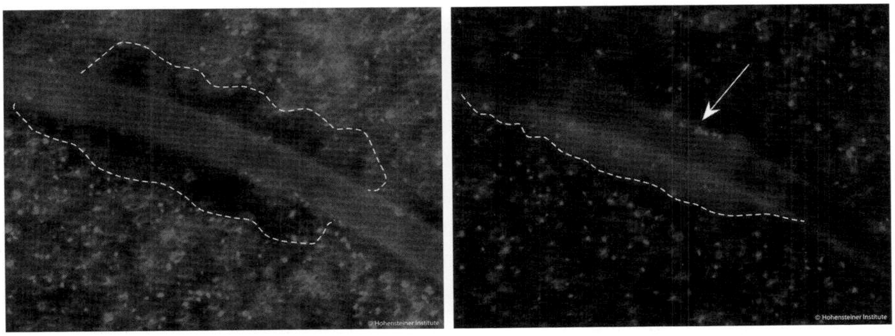 *b*

Fig. 4. ***a*** Fluorescence image of living germs along an antimicrobial fibre. The dashed line indicates the active zone of inhibition (50–100 μm), which lacks living micro-organisms. ***b*** Fluorescence image of dead germs along an antimicrobial fibre. Please note the attachment of dead germs on the fibre surface (arrow) and their accumulation within the zone of inhibition (dashed area).

same time, in the same testing procedure, use was made of a further fluorescent dye (propidium iodide), which is only absorbed by bacteria which have been killed (fig. 4b). Here, the same fibre displays red-fluorescent dead bacteria found in direct contact with the fibre or predominantly within the inhibition zone.

To summarize, the in vitro results can be interpreted as apparently indicating that germs are killed off only in very close and direct contact with fibres treated with an antimicrobial agent. Furthermore, bacteria are found especially in those parts of the textile where there are nutrients attached (e.g. skin flakes, foodstuffs, fats). These data suggest that the human skin flora can only be influenced where the skin is in direct and permanent contact with the treated fibre. Extensive wear trials with antimicrobial textiles show that there is a clear odour-reducing effect of the textile, but not the skin of the wearer. This is mainly due to the passive soaking effect, attaching sweat, skin germs and their substrates into the fabric and onto the active fibre.

Conclusion and Outlook

In hospitals in particular, textiles (gowns, smocks) along with the standard channels of transfer (hand-carried and droplet-borne infections) play a considerable role in the spread of pathogenic germs. It has been proven that, under certain circumstances, pathogenic airborne germs adhere to the textile and can in this way be transferred from ward to ward [18]. At present, there is an estimated number of up to 40,000 deaths occurring every year during a stay in a German hospital due to the patient acquiring some bacterial (nosocomial) infection. Apart from the protection of the patients and staff in the hospital, it should also be determined whether textiles treated with an antimicrobial agent also offer a greater degree of hygiene safety for wearers of personal protective equipment in the emergency services, foodstuff operations and the military environment. Interestingly, a current, previously unpublished wearing trial study carried out among hospital staff indicates that in a direct comparison experiment with special smocks which had been treated with an antimicrobial on the right side versus none on the left, after work the smock side with the treated fibres was distinctly less germ laden than the side which was not treated. Against this background, the use of antimicrobial clothing in certain areas in hospitals would seem to be thoroughly recommended for the prevention of infection.

Unlike the situation with outer protective clothing, for the reduction of odour in the sporting, leisure and private areas, textiles treated with antimicrobial agents tend to be worn close to the body. If one considers that depending on the design and type of fibre, only a few textile fibres are in direct contact with the skin, and then only briefly, the results presented in this article show that no dramatic change in skin flora should be anticipated. However, long-term wearing tests with treated textiles carried out under conditions of dermatological qualification should further underline the safety of the application.

Acknowledgement

The technical assistance of J. V. Seydlitz-Kurzbach and B. Gortan is gratefully acknowledged.

References

1 Lonsdale-Eccles A, Leonard N, Lawrence C: Axillary hyperhidrosis: eccrine or apocrine? Clin Exp Dermatol 2003;28:2–7.
2 Haustein U-F: Bakterielle Hautflora, Wirtsabwehr und Hautinfektionen. Dermatol Monatsschr 1999;175:665–680.
3 Boman HG: Antibacterial peptides: basic facts and emerging concepts. J Intern Med 2003;254: 197–215.
4 Gläser R, Harder J, Lange H, Bartels J, Christophers E, Schroeder JM: Antimicrobial psoriasin (S10047) protects human skin from *Escherichia coli* infection. Nat Immunol 2004;10:Ni1142.
5 Schroeder J-M: Epithelial antimicrobial peptides: innate local host response elements. Cell Mol Life Sci 1999;56:32–46.
6 Leyden JJ, McGinley KJ, Holzle E, Labows JN, Kligman AM: The microbiology of the human axilla and its relationship to axillary odor. J Invest Dermatol 1981;77:413–416.
7 Roth RR, James WD: Microbial ecology of the skin. Annu Rev Microbiol 1988;42:441–464.
8 Yamazaki K, Beauchamp GK, Singer A, Bard J, Boyse EA: Odor types, their origin and composition. PNAS 1999;96:1522–1525.
9 Kanda F, Yagi E, Fukuda M, Nakajiama K, Ohta T, Nakata O: Elucidation of chemical compounds responsible for foot malodour. Br J Dermatol 1990;122:771–776.
10 Guillet G, Zampetti A, Aballain-Colloc ML: Correlation between bacterial population and axillary and plantar bromidosis. Eur J Dermatol 2000;10:41–42.
11 Pelosi P: Odorant-binding proteins: structural aspects. Ann NY Acad Sci 1998;855:281–293.
12 Jacob S, McClintock MK, Zelano B, Ober C: Paternally inherited HLA alleles are associated with women's choice of male odor. Nat Genet 2002;30:175–179.
13 Rothardt G, Beier K: Peroxisomes in the apocrine sweat glands of the human axilla and their putative role in pheromone production. Cell Mol Life Sci 2001;58:1344–1349.
14 Gauger A, Mempel M, Schekatz A, Schafer T, Ring J, Abeck D: Silver-coated textiles reduce *Staphylococcus aureus* colonization in patients with atopic eczema. Dermatology 2003;207:15–21.
15 Abeck D, Mempel M: *Staphylococcus aureus* colonization in atopic dermatitis and its therapeutic implications. Br J Dermatol 1988;139:13–16.
16 JIS L 1902: Testing for antibacterial activity and efficacy on textile products, 2002.
17 Bartels VT, Umbach KH: Hautkontakt. Kettenwirkpraxis 2002;2:12–15.
18 Panknin G, Geldner G: Problemerreger auf Intensivpflegestationen unter Berücksichtigung von MRSA. Hartmann Wundforum 2000;4:15–21.

Dr. Dirk Höfer
Institute for Hygiene and Biotechnology, Hohenstein Research Center
DE–74357 Boennigheim (Germany)
Tel. +49 7357 271432, Fax +49 7357 271 91432, E-Mail d.hoefer@hohenstein.de

Burg G (ed): Biofunctional Textiles and the Skin.
Curr Probl Dermatol. Basel, Karger, 2006, vol 33, pp 78–109

Hygienic Relevance and Risk Assessment of Antimicrobial-Impregnated Textiles

A. Kramer[a], *P. Guggenbichler*[d], *P. Heldt*[a], *M. Jünger*[b], *A. Ladwig*[b], *H. Thierbach*[a], *U. Weber*[c], *G. Daeschlein*[a]

[a]Institute of Hygiene and Environmental Medicine and [b]Clinic of Dermatology, Ernst-Moritz Arndt University, [c]Hygiene Nord GmbH, Greifswald, [d]Department of Pediatrics, University of Erlangen-Nürnberg, Erlangen, Germany

Abstract

The antimicrobial impregnation of textiles is intended to provide protection of textiles against microbial corrosion, prevention of malodor or prophylaxis and therapy of infections, respectively. For every biocidal product a careful risk assessment for humans and the environment has to be performed. The advantage of antimicrobially active textiles has to be documented for every agent as well as for every application, and a balance has to be found between a textile's quality rating and the potential risks, e.g. sensitization, disturbance of the ecology of the skin, toxic side effects by means of systemic absorption, cytotoxicity, genotoxicity, carcinogenicity, teratogenicity and ecotoxicity. This article evaluates the applicability of silver compounds as well as the classic antimicrobials triclosan, quaternary ammonium compounds, copper and further new options like chitosan and zeolite. It has to be emphasized that there are no objections against the use of antimicrobially active textiles if their use is equal or superior to other preventive or therapeutic measures. This applies to the amelioration of the course of dermatological diseases with disturbed skin flora, in particular atopic dermatitis, the prevention and therapy of acute and chronic wound infections by wound dressings, the use of impregnated surgical suture material as well as special indications in the prevention of infection in medical facilities. The use of antimicrobial textiles for the prevention of dermatomycosis by antifungal impregnation is of questionable use; the antimicrobial impregnation of textiles for deodorization purposes has to be avoided. Presently, from a hygienic point of view, the following questions have to be clearly determined: declaration of any antimicrobial impregnation; development of international standards for in vitro testing and preclinical evaluation of efficacy and tolerance; evaluation of the advantage of the antimicrobial properties for the intended use including the risk-benefit assessment.

Herbs and spices for the conservation of textiles, e.g. for embalming mummies, were already used by the pharaohs [Vigo, 1983]. During the

Second World War, the German army used uniforms impregnated with quaternary ammonium compounds (quats) under the suggestion that secondary wound infections can be prevented [Arnold, 1963]. At the same time Engel and Gump [1941] recommended hexachlorophene impregnation of cotton and wool.

With the worldwide expansion of synthetic textiles the need for antimicrobial impregnation of textiles increased substantially as synthetic textiles absorb about 25% less water vapor compared to cotton or wool. The blocking of evaporation of sweat results in a thin fluid film on the skin, which consequently impairs further the evaporation of sweat, providing an ideal environment for proliferation of bacteria and fungi. Soon, commercial interest was directed to the antimicrobial impregnation of natural fibers.

Purpose and Risks of Antimicrobial-Impregnated Textiles

The purpose of antimicrobial impregnation of textiles is:

- elimination of malodor by deodorization or absorption;
- increasing the lifetime of technical textiles, but also of household textiles and leather by prevention of microbial corrosion (e.g. moulding, discoloration);

an official assessment of 109 textile and leather samples disclosed biocidal concentrations between 1 and 649 mg/kg in 77 samples, thereof 4-chlor-3-cresol up to 200 mg/kg, 2-phenylphenol up to 210 mg/kg and 2-(thiocyanatomethylthio) benzothiazole up to 540 mg/kg in leather clothing, and pentachlorphenol(!), 4-chlor-3-cresol and 2-phenylphenol up to 2 mg/kg and triclosan up to 640 mg/kg in textiles (www.bfr.bund.de, 11th session of the working group 'health assessment of textile adjuvants and dyes');

- prevention of dermatomycosis by antifungal impregnation;
- impact on the progress of dermatological diseases by alteration of skin flora;
- prevention or therapy of wound infections;
- prevention of infections in medical facilities.

Regarding the side effects for humans and the environment, every textile with antimicrobial activity must undergo a risk assessment. At the same time, also the benefits derived from the use of the biocide have to be assessed and balanced for its risks.

The following side effects can be provoked by biocides:

- sensitization and manifestation of allergic disorders, in particular as contact eczema;

- anaphylactic reaction (presently described only for chlorhexidine [Kramer, 2001]);
- negative impact on the microecology of the skin with influence on colonization resistance (presently not evaluated in detail);
- induction of development of antimicrobial resistance including cross-resistance with antibiotics;
- with long-term application dermal absorption of biocides and dose-related chronic toxicity including genotoxicity, cancerogenicity and teratogenicity have to be investigated;
- in case of missing or low-degree biological degradation, of high-level burden or toxification of biocides, the cumulative ecotoxic effects have to be assessed.

In Germany the actual federal regulation of textile declaration requires only information about the textile fibers but not about the adjuvants. However, according to the food and legislation act for articles of daily use (§ 30), the manufacture of articles of daily use with potential health hazards is restricted. Up to now this has required neither declaration nor obligatory registration. The presently used practice to declare biocides only in case of allergic potential is not satisfactory. Regarding the risk potential a general declaration is required. In future, biocides for impregnation of textiles can only be used, when they are contained in the 'positive list' of the biocide product guideline.

Hygiene Assessment for the Use of Antimicrobial-Impregnated Textiles

Deodorants

Besides local dermal applications impregnated textiles or synthetic fleece are rarely used for deodorization [Untiedt, 2004]. The intended use for textiles impregnated with a deodorant is to prevent bacterial degradation of sweat not at the site of the skin but in the textile itself. The impregnation of textiles for deodorization is inevitably less active than the direct application of the deodorant to the skin, as the contact of an antimicrobial-impregnated textile with the skin, e.g. in the axilla, is substantially worse. Therefore this application cannot legitimately be recommended. Regarding the environmental burden, the application at the site of the textile has to be considered as more critical than the dermal application.

The use of antibacterial textiles for infants and children is controversial. In the light of the hypothesis that infections in early childhood may be protective against the development of allergic disorders [Strachan, 1989; Cookson and Moffatt, 1997; Matricardi, 1997; Matricardi and Bonini, 2000; Matricardi et al., 2000; Friedrich et al., 2006], reduction of the skin flora might be hazardous [Bodner et al., 1998; Farooqi and Hopkin, 1998; Bager et al., 2002; Gibbs et al., 2004].

A potential explanation provides the Th1/Th2 paradigm [Mosmann et al., 1989; Romagnani, 1992]. The two helper cell subsets Th1 and Th2 are characterized by different patterns of cytokine production and have different functions. Th2-derived cytokines inhibit the development of Th1 cells and vice versa. Atopic diseases are Th2 mediated and characterized by the release of IgE, whereas bacterial and viral infection are more likely to be Th1 mediated. Thus, the infection-induced Th1-cell-specific cytokine inhibits the development of allergen-specific Th2 cells. This theory is however not uniformly accepted.

The endowment of towels with deodorizing agents is principally contradictory. This is also valable for the addition of deodorants into the wash process for laundry and rugs.

Conclusion: Dirty and smelling textiles must be cleaned with detergents. It does not make sense to block and to override the smell by deodorants.

Protection against Mouldiness and Bacterial Destruction

The antimicrobial impregnation of textiles improves durability. Antimicrobial textiles can be used for tents, covers, curtains, nets, sacks, technical felts, filters and special work clothes [Wallhäusser and Fischer, 1970]. The choice of the appropriate antimicrobial agents must follow strict toxicological and ecological criteria. The necessity must be carefully evaluated.

Antifungal Impregnation

Modern clothing materials provide an optimal milieu for growth of bacteria and fungi, because synthetic fibers create a moist chamber with maceration of the skin [Meyer-Rohn and Kulenkamp, 1975]. As synthetic fibers cannot be boiled, fungi and bacteria can survive and cause colonization, infection or reinfection, respectively. Shoes play a crucial role in the development of plantar mycosis. In a study with several thousands of soldiers, it was found that wearing sandals resulted in 3.5% plantar mycosis in contrast to 28% of persons wearing closed shoes [Taplin, 1976]. In common public facilities [Salminen et al., 1974; Kraus and Tiefenbrunner, 1975] as well as in pedicure salons, shoe shops and locker rooms, fungal spread takes place, which cannot be controlled by the use of antifungal textiles. This can predominantly be achieved by personal hygienic measures, wearing comfortable well-aerated shoes, avoidance of barefoot walking in risk zones and antimycotic therapy in early stages of infection. In public facilities with high risks for dermatomycosis, basic preventive measures by the employer are mandatory. Carefully drying the feet including the space between the toes together with skin care is important [Seebacher and Kramer, 1997].

Conclusion: There is neither an epidemiological nor a hypothetic justification for antifungal endowment of textiles for the prevention of fungal infections.

It is rather sufficient to change textiles regularly, to wash or clean underwear and socks at a minimum of 60°C. If this is not possible, the laundry of persons with overt dermatomycosis as well as of particularly predisposed persons can be washed at 30°C, yet the addition of disinfectant cleaner is helpful.

Adjuvant Therapy of Dermatomycoses

In contrast to the previous opinion that an allergic pathogenesis is exclusively responsible for the typical lesions seen in patients with atopic dermatitis, recent research results suggest a 2-step pathogenesis. It is still commonly accepted that the initial trigger is an allergic antigen-antibody (IgE) reaction. However, newer studies found that bacteria and fungi are responsible for the maintenance of clinical symptoms, for the exacerbations and progression of the disease. The skin, altered by the inflammatory process, is colonized by substantially higher amounts of bacteria. Therefore modern therapeutic strategies focus on the adjuvant antimicrobial therapy of atopic dermatitis.

An early start of therapy with antimicrobial active immune modulators, e.g. tacrolimus and pimecrolimus, which are thought to modify the allergic part of the disease, can prevent the outbreak and the spread of eczematous skin lesions. In addition to the immunomodulatory effect, these medications are antimicrobially active as the basic molecule of these compounds is a macrolide antibiotic.

A good clinical efficacy has been demonstrated with silver-coated antimicrobially active textiles without evidence of bacterial resistance. It has been documented that the antiseptic activity of silver-coated textiles can reduce the colonization with *Staphylococcus aureus* as well as the production of staphylococcal exotoxin whereby the inflammatory reaction is suppressed [Werfel, 2001]. In contrast to impregnation of textiles with silver salts, metallic silver exhibits a good tissue tolerance. Starting with an enteral absorption of >2 g of silver, argyrosis has been observed, which is a deposition of silver in the superficial skin layers. The amount of silver, which is absorbed even through damaged skin by the inflammatory process, is 1,000-fold below the toxic level even when the textile is used over years.

The cumulative costs of therapy are limited to the initial costs of the silver-coated textile.

Nevertheless the declaration of silver impregnation cannot be correlated with clinical efficacy. Three different textiles endowed with silver by different technologies (No. 1: 50% of fibers are coated with metallic silver, No. 2: 50% polyamide fibers with a durable silver staining, No. 3: silver-impregnated Trevira) have been investigated for their antimicrobial activity.

Pieces of 1 cm^2 of the textiles were incubated with a suspension of 10^9 colony-forming units (CFU) of the test organism. In 3-hourly intervals, 50 μl of

Table 1. Number of surviving bacteria after exposure to three silver textiles (No. 1–3) for various reaction times

Textile No.	Test organism	Bacterial count, log CFU						
		3h	6h	9h	12h	15h	18h	24h
1	*S. aureus*	9	8	6	4	0	0	0
	S. epidermidis	8	6	4	0	0	0	0
	E. coli	6	4	0	0	0	0	0
	P. aeruginosa	8	7	6	5	0	0	0
2	*S. aureus*	9	8	6	5	0	0	0
	S. epidermidis	8	6	0	0	0	0	0
	E. coli	6	4	0	0	0	0	0
	P. aeruginosa	8	7	5	0	0	0	0
3	*S. aureus*	9	9	8	8	7	7	7
	S. epidermidis	9	8	8	7	6	6	6
	E. coli	9	8	7	7	7	6	5
	P. aeruginosa	9	8	8	8	7	7	6

The value 0 means below the level of detection of 10^5 CFU/ml.

the suspension were inoculated on Columbia agar (containing 50% sheep blood). After incubation for 24 h the number of CFU was determined.

Table 1 shows substantial differences between the different silver textiles [Guggenbichler, unpubl.].

In another study the influence of a moisture-permeable silver textile on the skin of 5 healthy volunteers was investigated.

An area of 10 × 10 cm was tightly covered with one of the samples manufactured according to the technologies described above . Every 3 h for the first 9 h and 6 h thereafter the number of CFU on 1 cm^2 of skin was determined.

At the beginning of the experiment, a mixed flora consisting of *Staphylococcus epidermidis, Propionibacterium acnes* and *Streptococcus mitis* in an amount of 4.3–5.5 log CFU/cm^2 was isolated. Textiles No. 1 and 2 reduced the skin flora every 3 h by 1 log, so that at 9 h no organisms could be isolated any more. This effect lasted for another 9 h; thereafter, the normal flora gradually recolonized the skin so that at 36 h the normal skin flora in the amount mentioned before could be detected. No irritation of the skin occurred during the experiment. Textile No. 3 did not show any antibacterial effect. Twenty-four-hour extracts of these three textiles in physiological saline were investigated for cytotoxicity on mouse fibroblasts. No reduction of viability was seen with the methylene blue test with any of the textiles [Guggenbichler, unpubl.].

Table 2. Mean total number ± SD of isolated microorganisms in neurodermitic skin lesions (log CFU/cm^2) of patients (determination by direct agar contact method)

Group	Day 0	Day 14	Day 28	p^1	Day 56	p^2	p^3
Ag textile	1.87 ± 0.51	1.41 ± 0.82	1.35 ± 0.64	0.07	1.45 ± 0.46	0.6	0.03
Cotton	1.59 ± 0.88	1.41 ± 0.97	1.42 ± 0.67	0.6	1.53 ± 0.95	0.9	0.9

[1]t test with paired samples, comparison between day 0 and day 28.
[2]Comparison between day 28 and day 56.
[3]Comparison between day 0 and day 56.

In a controlled clinical trial with patients suffering from atopic dermatitis, the influence of silver underwear on the normal skin flora was investigated over a period of 1 month. Patients in the first group were wearing silver textiles manufactured according to technology No. 1 (see above) throughout from day 0 to day 28. The patients in a second group were wearing identical cotton underwear without silver from day 0 to day 14 and then the silver textile from day 14 to day 28. From day 28 all patients returned to their individual underwear. In the first group bacterial reduction on the skin with neurodermitic lesions over a period of 56 days was observed. In the group with the patients wearing silver textiles for 14 days, no significant reduction of the skin flora was noted (table 2). Apparently wearing this silver underwear for 14 days is not sufficient for a substantial reduction of the skin flora. The analysis of microorganisms revealed an analogous reduction for *S. aureus* and *S. epidermidis*. The lower number of colonizing organisms which was observed on day 56 warrants additional attention as it is assumed to be a remanent effect. It is conceivable that silver deposits in the skin could be responsible for this reduction. In order to obtain further insight into this hypothesis, silver elimination in the urine was investigated. In the urine no silver was detected (threshold of detection 0.9 μg/ml).

Clinical evaluation of the patients revealed that silver underwear worn over the longer period resulted in a significant improvement of the SCORAD (Scoring Atopic Dermatitis) index compared to the group with a shorter wearing period. The silver underwear does not tolerate washing temperatures above 60°C. Therefore the overall hygienic impression was unsatisfactory. After wearing the textiles for more than 2 days, the patients were disturbed by malodor and discomfort, which increased with duration of use [Jünger et al., 2006].

Other groups of investigators achieved comparable results. In an open controlled trial with 15 patients, two sites of neurodermitic lesions in the same patients were compared over 14 days. The elbow of one arm of the patients with neurodermitis was covered for 7 days with a silver-coated material, the other site with normal cotton. Thereafter an observation period of 7 days followed. After 2 days a significant reduction of colonization with *S. aureus* was observed on the site covered with the silver textile. The total number of CFU of skin flora remained below the initial concentrations of skin microorganisms on the control site for another 7 days. The SCORAD at the beginning of the experiment was identical on both arms. On days 2, 7 and 14 the SCORAD was substantially improved on the silver-textile-covered arm. During the follow-up period, the difference between the silver textile and the control site was still accentuated. In 5 patients the condition remained stable; in another 6 patients further improvement was observed. At the site covered by cotton without silver, a slight but nonsignificant clinical improvement of the eczematous lesion was observed on day 2. On days 7 and 14 the SCORAD deteriorated. The comparison between the two sites revealed an optical improvement with the silver textile ($p < 0.05$) on days 7 and 14 [Gauger et al., 2003].

In a randomized, controlled clinical trial with atopic dermatitis patients (35 patients with silver-coated textiles, 22 with placebo), an improvement of the SCORAD – in the silver group of 27.4% ($p < 0.05$) and in the placebo group of 16.3% ($p > 0.05$) – was observed. The affected area of the skin in the silver group regressed by 16.6% ($p < 0.05$) and by 8.3% ($p > 0.05$) in the control group; the difference versus placebo was at the threshold of significance ($p = 0.051$). An equal improvement was seen in the DIELH score (Deutsches Instrument zur Erfassung der Lebensqualität bei Hauterkrankungen; score of Schäfer et al. [2001]). After 7 days, the patients noted a recognizable improvement in itching, comfort and dryness in the silver versus the placebo group ($p < 0.05$). The differences became accentuated after 14 days. Also, concomitant medication (corticosteroid ointments) was more frequent in the placebo group ($p > 0.05$) [Schäfer, in prep.].

Conclusion: An improvement in the therapy of atopic dermatitis by silver-coated/impregnated textiles could be demonstrated and must be confirmed in larger studies, before being advised as therapeutical option.

Prevention of Postoperative Wound Infections

The impregnation of suture material is a promising approach to the prophylaxis of postoperative wound infections.

After surgical interventions in microbially colonized areas, the risk of wound dehiscence is given. Interventions in areas colonized with $<10^5$ CFU/g tissue normally are without risk for postoperative wound infections in immunocompetent hosts [Elek and Conen, 1957]. However, with impaired host defense mechanisms, foreign bodies and ischemia, even substantially lower amounts of microorganisms ($<10^3$ CFU) are able to cause a wound infection [Schmitt, 1991]. In the presence of suture material even concentrations of 10^2 *S. aureus*/g tissue are critical [Elek and Conen, 1957]. Sutures in abdominal surgery are generally pulled through the lumen of the gut and are contaminated with a variety of aerobic and anaerobic bacteria. Under these circumstances a wound infection and dehiscence of the suture must be taken into consideration. Clinical experience shows that sutures in the large intestine are more frequently insufficient than sutures in the stomach. The proportion by which sutures are responsible for a primary wound infection can only be resolved by clinical studies comparing the incidence of postoperative wound infections in patients with antimicrobial sutures and in controls. Vincent [2003] reports an incidence of postoperative wound infections of 15%. The risk of implanted foreign bodies, e.g. sutures or clips as primary cause for nosocomial wound infections, is estimated to be low; however, implanted biomaterials play a major role in the perpetuation of infections; the percentage is unknown (no studies).

In vitro studies demonstrate a zone of inhibition for *S. epidermidis* around a triclosan-impregnated suture of $14.5\,cm^3$, for *S. aureus* or methicillin-resistant *S. aureus* (MRSA) of $17.8\,cm^3$. The antibacterial activity lasted 7 days [Rothenburger et al., 2002]. However, a disadvantage of triclosan-impregnated sutures is the intrinsic resistance of triclosan against *Pseudomonas aeruginosa, Serratia marcescens* and *Alcaligenes* spp. In vivo studies confirmed the efficacy.

Sutures were implanted in the dorsolateral thighs of guinea pigs; 5×10^4 CFU *S. aureus* were instilled into the wound by a catheter, and 48 h later the suture was explanted. Thereafter $10^{3.6}$ CFU were isolated from nonantimicrobially active suture material in contrast to $10^{1.85}$ CFU in triclosan-impregnated sutures ($p < 0.05$) [Storch et al., 2002a, 2004]. The addition of triclosan to the suture did not influence physical properties and the handling of the device [Storch et al., 2002b].

No cytotoxic effect of the impregnated sutures was observed in tissue cultures, and the material was nonpyrogenic. No intracutaneous and intramuscular side effects were observed. Between the triclosan-impregnated materials and nonimpregnated materials, in none of the animal models differences could be detected [Barbolt, 2002]. Accordingly any difference in the healing process of experimental wounds in guinea pigs could be disclosed [Storch et al., 2002a].

Conclusion: The use of antimicrobial-impregnated surgical suture materials is beneficial, particularly in critically contaminated wounds and in patients with a high risk of infection. The benefit for reduction of postoperative wound infections has to be documented in clinical trials. The uptake of triclosan with a suture is toxicologically noncritical. As long as the in vitro inducible development of resistance to triclosan has no clinical relevance, the use in suture material is uncritical [Kramer et al., 2005a]. A promising alternative approach will be the use of nanocristalline silver.

Table 3. IC_{50} of antiseptic agents on mouse fibroblasts after 30 min of exposure [Müller and Kramer, unpubl.]

Agent	IC_{50}, mg/ml
Ag[1]	0.007–0.055
Silver sulfadiazine	0.030–0.040
Polyhexanide	0.150–0.200
Povidone-iodine	4.7–4.8

[1]Results of Poon and Burd [2004], calculated on the basis of silver.

Antiseptic Wound Dressings

For the prevention of postoperative and posttraumatic wound infections as well as for the treatment of acute wound infections, antisepsis (irrigation, ointment, gel) in combination with surgical debridement is the method of choice [Kramer et al., 2004]. For colonized or infected chronic wounds, it is important to eliminate the microorganisms and toxins without disturbing the wound-healing process. For this purpose antiseptic wound dressings with simultaneous absorption of toxins are useful [Müller et al., 2003]. Presently silver-containing wound dressings represent the state of the art. Wound dressings which do not release silver into the tissues are preferable especially to prevent disturbance of the healing process.

The toxic effect of Ag^+ is caused by the interaction of silver ions with the cell membrane and the respiratory chain (reaction with cytochromes b and d and the substrates of oxidation [Weber and Rutala, 2001]) of newly formed epithelial cells. In keratinocyte and fibroblast cell cultures, $AgNO_3$ and nanocristalline silver are highly toxic in concentrations of 7×10^{-4} to 55×10^{-4}% [Poon and Burd, 2004]. In comparison povidone-iodine and polyhexanide show substantially less cytotoxicity (table 3).

Hidalgo et al. [1998] investigated the therapeutic activity (MIC/cytotoxicity) of silver nitrate. Depending on microorganisms, growth inhibition was detected after 100- to 700-fold dilution with even measurable cytotoxicity. As a consequence 0.01% was recommended as therapeutic concentration. Innes et al. [2001] compared the time for reepithelialization in a mesh graft model (prospective controlled matched-pair study). The criterion of >90% reepithelialization ($p = 0.004$) was seen in a silver-free dressing environment after 9.1 ± 1.6 days, in a silver-containing dressing after 14.5 ± 6.7 days. There was no difference in bacteriological culture positivity. After 1 and 2 months, scar formation was significantly worse by use of a silver-containing dressing compared to a silver-free dressing. After 3 months the initial differences evened out. In addition the risk of systemic side effects by liberation of free silver ions has to be taken into consideration. After the application of silver sulfadiazine on burn wounds, silver concentrations of 440 μg Ag/l blood and 12 μg Ag/l urine were found [Maitre et al., 2002]. The authors recommend the monitoring of silver levels, when silver-releasing compounds are used.

In order to estimate the relevance of the in vitro cytotoxicity, we evaluated the minimal microbicidal concentration (MMC, defined as reduction factor ≥5 log steps) in 10% fetal calf serum in a quantitative suspension test. Silver sulfadiazine was ineffective. 1% $AgNO_3$ reduced the inoculum size of *Escherichia coli* by maximally 1 log. Within 30 min all silver-containing compounds were inactive against *S. aureus*. Against *E. coli* the concentration of >5% of mild silver protein solution was active. For the comparison of antiseptics we introduced a biocompatibility index as quotient of IC_{50} and MMC ≥5 log. For $AgNO_3$ and silver sulfadiazine no biocompatibility index could be calculated, because the highest tolerated concentration (1%) did not result in any reduction of the initial inoculum size. Colloidal silver resulted in a biocompatibility index (L929-cells/*E. coli*) of 0.128, chlorhexidine had one of 0.729, povidone-iodine (related to iodine) one of 0.95, polyhexanide one of 1.325 and octenidine one of 1.506. Silver-containing antiseptics are highly cytotoxic without measurable antimicrobial activity within 30 min [Müller et al., 2006]. However, after 3 h of exposure of a wound dressing, a reduction of 5 log steps against *P. aeruginosa* ATCC 15442 was noticed [Müller et al., 2003]. These data indicate that a slow release of low silver ion concentrations is effective after an extended period of time. The induction of silver resistance cannot be excluded in case of prolonged and low-level release of silver ions [Percival et al., 2005].

Conclusion: Colonized or infected chronic wounds require the application of adequate antiseptic wound dressings with the ability of toxin absorption, e.g. endotoxins. When silver-containing wound dressings are used, the liberation of silver into the wound environment should not reach cytotoxic concentrations. The antimicrobial activity of silver in dressings cannot be compared with the rapid bactericidal action achieved by other antiseptics. Over a period of several hours, low nontoxic concentrations of silver ions are antiseptically effective and block the attachment of microorganisms to epithelial cells as well as biofilm formation. The activity of silver in dressings depends on the technology of silver bonding.

Prevention of Infections in Medical Facilities

No evidence exists for any preventive effect of antimicrobial-impregnated bedlinens, diapers, work clothes and towels. If basic hygienic measures are taken (e.g. hand disinfection, disinfection of patients near surfaces, change of textiles at regular intervals, barrier nursing) the antimicrobial equipment of textiles cannot provide additional protective effects. The skin as reservoir of microorganisms remains active as a source of bacterial spread; nevertheless, the amount of bacteria can be reduced. In addition the risk of sensitization has to be taken into consideration. In contrast, the antimicrobial endowment of air filters can be sensible, if no active compounds are liberated into the environment. Equally the antimicrobial endowment of lacquers, e.g. for door knobs or siphons, of flexible partition walls and of textile linings is meaningful and allows the avoidance of prophylactic disinfection. For this purpose nanocristalline silver seems to be the agent of choice.

The disinfection of surfaces or skin antisepsis by means of fleeces impregnated with biocide must be considered as a special case of application of antibacterial textiles.

In Germany the solution for fleeces is listed in the positive list of the German Confederation of Applied Hygiene, if the recommended antimicrobial activity of the solution is fulfilled. This does not apply to antimicrobial-impregnated disinfecting fleece and requires a comment insofar as the impregnated fleece does not achieve the activity of the active solution per se. One textile, impregnated with 70% propan-2-ol, is effective and registered as drug for skin antisepsis in Germany.

Conclusion: For special indications antimicrobial-impregnated textiles can provide a valuable contribution to the prevention of infection.

Methods of Antimicrobial Impregnation of Textiles and Hygienic Consequences

Highly diffusible impregnations are differentiated from nondiffusible or poorly diffusible antimicrobial impregnations. For impregnation, the Foulard application is most often used, where the active compound is applied during the spinning process. Further techniques for application of antimicrobial substances are coating, vapor deposition (e.g. silver), spray application, aftertreatment in a bath (e.g. for triclosan) or treatment of prefabricated items in a washing machine, which is ecotoxicologically questionable. Even when liberation into the environment is excluded, the ecological safety of the applied biocide with respect to waste disposal must be considered. Besides, the impossibility of liberation of the active agent must de documented not only for the ready-to-use product, but also during usage, because sweat, antimicrobial degradation and/or the washing process can lead to hazardous secondary products. It has been documented that triclosan can be transformed into the toxic 2,7/2,8-dibenzodichloro-*p*-dioxin by sunlight in sewage; this can be detected in sewage plants [Mezcua et al., 2004].

Characteristics of Selected Compounds for the Antimicrobial Impregnation of Textiles

Taking triclosan as an example, the problems of antimicrobial impregnation of textiles will be demonstrated. Other compounds are described briefly.

Triclosan (2,4,4'-Trichloro-2'-Hydroxydiphenylether)
Antimicrobial Activity and Spectrum of Activity

Within 10 min of exposure, the MMC measured for *S. aureus* and *Candida albicans* was 25 μg/ml, for *E. coli* 500 μg/ml. Within 72 h the MMC for *S. aureus* was 0.1 μg/ml, for *E. coli, Proteus* spp. and *Klebsiella pneumoniae* 0.03–0.3 μg/ml, for *Enterobacter aerogenes* 1–3 μg/ml, for *Enterococcus faecalis, C. albicans* and *Saccharomyces cerevisiae* 3–10 μg/ml [Räuchle, 1987; Wallhäusser, 1995]. *P. aeruginosa* shows a high intrinsic resistance with an MIC of >1,000 μg/ml [Räuchle, 1987; Chuanchuen et al., 2003]. Triclosan inhibits the growth of *Pityriasis versicolor* at a concentration of 75 μg/ml within 1–2 days almost completely but does not eradicate the organism [Hundt et al., 2000].

The mode of action consists in the inhibition of the enoyl-acyl carrier protein reductase [McMurry et al., 1998, 1999; Levy et al., 1999; Heath and Rock, 2000], which destabilizes the membrane. The fatty acid synthase system (type II) of bacteria is substantially different from the mammalian enzyme, which can be an option for selective inhibition [Marrakchi et al., 2002].

Development of Resistance

Exposure of microorganisms to subbactericidal concentrations in vitro can induce resistance [Hundt et al., 2000; Hundt, 2001; Brenwald and Fraise, 2003]. Cookson et al. [1991] described strains of MRSA with MICs between 2 and 4 μg/ml from patients with daily triclosan total body washings, whereas *S. aureus* strains from patients without exposure to triclosan exhibited MICs between 0.01 and 0.1 μg/ml. As a result of the specific point of attack of triclosan on the bacterial cell, the enoyl-acyl carrier protein reductase [Heath and Rock, 2000], and because the resistance development mechanisms against antibiotics are comparable to that against triclosan (target mutation, increased target expression, active efflux from the cell, enzymatic deactivation/degradation), laboratory findings related to the cross-resistance between triclosan and antibiotics are not surprising [Russell et al., 1998; Chuanchuen et al., 2001, 2002, 2003; Braoudaki and Hilton, 2004, 2005; Randall et al., 2004; Sanchez et al., 2005].

The genetic product of *fab*I, the enoyl-acyl carrier protein reductase, evolved as a major target site for the development of resistance to triclosan [McMurry et al., 1999]. *E. coli* strains with mutation of the *fab*I gene frequently showed low, medium or high triclosan resistance (MICs between 0.2 and

25 μg/ml [McMurry et al. 1998]). *S. aureus fab*I mutants also showed increasing MICs (1–4 μg/ml) against triclosan [Brenwald and Fraise, 2003]. *Mycobacterium smegmatis* strains with a mutation of the *inh*A gene are resistant to triclosan as well as to isoniazid. It has been suggested that also the tuberculostatic activity of isoniazid is based on *inh*A products [Levy, 2001]. The discussion whether plasmid-coded antibiotic resistance is able to induce triclosan resistance is still controversial. Although it has been shown by Cookson et al. [1991] that triclosan resistance against *S. aureus* can be transferred simultaneously with the plasmid-coded mupirocin resistance, this could not be verified by other investigators [Suller and Russell, 2000].

On the basis of these laboratory findings it would be conceivable that, as result of the widespread use of triclosan particularly in consumer products [Furuichi et al., 1999; Braid and Wale, 2002] antibiotic resistances could be selected [Schweizer, 2001], although up to now no organism with acquired triclosan resistance has been described [Gilbert and McBain, 2002; Screenivasan and Gaffar, 2002; De Vizio and Davies, 2004]. Also the MIC values have remained stable during the last decade [Goodfellow et al., 2003]. No clinical data exist regarding a triclosan-induced resistance against antibiotics [Suller and Russell, 1999; Russell, 2000, 2002, 2003]. The clinical relevance of the previous laboratory findings has to be elucidated [Suller and Russell, 2000]. Until this point the use of triclosan should be limited to medically proven indications.

Acute Toxicity

On the basis of LD_{50} (mg/kg body weight, BW), determined by oral application, to mice of 4,500, to rats of 3,700–5,000, to dogs of 5,000 and, by dermal application, to rabbits of 9,300 [Räuchle, 1987], the compound is classified as 'little to nontoxic'. The subcutaneous LD_{50} for rats is $>$147,000 mg/kg BW [De Salva et al., 1989], which suggests the categorization 'nontoxic'.

In the rat, no changes of glutamic oxaloacetic transaminase, glutamic pyruvic transaminase and blood urea nitrogen were observed after a single oral dose of 625 and 2,500 mg triclosan/kg BW in contrast to chlorhexidine after oral administration of $\geq$1,000 mg/kg BW. In vitro, a dose-related inhibition of the accumulation of *p*-aminohippurate but not of N-methylnicotinamide in the kidney of male rats was observed. The clinical relevance of these laboratory findings has to be established [Chow et al., 1977].

Skin Tolerance and Photosensitization

The antiseptic concentration for use was tolerated without adverse effects. A good skin tolerance has even been established in the 'repeated-insult patch test' with soaps containing 10% triclosan [Räuchle, 1987]. Trials with humans did not reveal any phototoxic reactions [Kligman and Breit, 1968].

Eye Tolerance

Solutions of 1–10% induced a transient hyperemia and chemosis which resolved after 24 h [Räuchle, 1987].

Sensitization and Photosensitization

Photosensitization was demonstrated neither in animal experiments nor in clinical studies [Marzulli and Maibach, 1973; Thomann and Maurer, 1975; Räuchle, 1987; Wnorowski, 1994]. Although occasionally photosensitization has been described, the widespread use of triclosan in deodorants and soaps apparently indicates an exceedingly low potential for sensitization [Roed-Petersen et al., 1975; Lachapelle and Tennstedt, 1979; Hindson, 1975; Wahlberg, 1976; Veronesi et al., 1986; Steinkjer and Braathen, 1988; Wong and Beck, 2001; Gloor et al., 2002]. However, 10 from 88 patients with (photo)allergies to UV filters in suncreams reacted to triclosan [Schauder, 2001]. In a second investigation among 103 patients, 3 responded with an allergic contact reaction and none with a photoallergic reaction [Steinkjer and Braathen, 1988]. Two of these patients received ointments containing corticosteroid and 3% triclosan. A Swedish investigation showed a prevalence of 0.2% (1,100 patients investigated) of contact allergy to triclosan [Wahlberg, 1976].

Subacute Toxicity

The subacute toxicity test (28 days) in monkeys revealed a no observable effect level (NOEL) of 100 mg/kg BW/day [Räuchle, 1987]; the daily administration of 0.1% triclosan solution over 3 weeks to dogs revealed no signs of toxicity [Schmid et al., 1994].

Subchronic Toxicity

In a subchronic toxicity test (90 days of oral administration for various animal species), the NOEL (mg/kg BW/day) reached for hamsters was 75, for rats 50, monkeys 30, dogs 12.5 and rabbits 3 [Paterson, 1969; Goldsmith, 1983; Räuchle, 1987; Schmid et al., 1994]. With dermal application of 3% triclosan in propylene glycol to rabbits, no local or toxic reaction was observed [Räuchle, 1987].

Chronic Toxicity and Cancerogenicity

For monkeys the NOEL was 30 mg/kg BW after oral administration over 1 year [Drake, 1975]. In a 2-year test in rats, 250 and 750 mg/kg in nutrition were well tolerated. After administration of 2,200 mg/kg, a mild reversible liver hypertrophy developed [Räuchle, 1987].

The inducible hepatotoxicity after administration of high doses of triclosan is obviously due to a competitive and noncompetitive inhibition of the 3-methylcholanthrene- and phenobarbital-inducible P450-dependent monooxygenase in liver microsomes by triclosan [Hanioka et al., 1996].

A feeding test over 2 years revealed no signs of cancerogenicity at a doses of 168 mg/kg/day in male rats or 218 mg/kg/day in female rats [Yau and Green, 1986; Räuchle, 1987]. The serum concentrations of triclosan in this test were between 26 and 27 mg/l. An independent review of this study reaches the same conclusion [Goodman, 1990]. Also in hamsters the harmlessness was documented [Chambers, 1999]. A dermal application of 0.5 and 1% triclosan in acetone over 18 months was well tolerated, and no signs of cancerogenicity occurred [Räuchle, 1987].

Mutagenesis and Reproduction Toxicity

Neither in vitro nor animal experiments revealed any signs of mutagenesis, embryotoxicity or teratogenic effects [Russell and Montgomery, 1980; Gocke et al., 1981; Räuchle, 1987; Henderson et al., 1988a, b; Jones and Wilson, 1988; Morseth, 1988; Riach et al., 1988; Denning et al., 1992; Schroder and Daly, 1992].

Absorption and Elimination

In experiments with rats, rabbits and dogs no organospecific accumulation of triclosan was detected. The elimination occurs after conjugation with glucuronic acid in the feces, in rabbits mainly renally. In humans the renal elimination as glucuronic acid or as sulfate conjugate represents the main route. The half-life counts 10 days without signs of cumulation [Räuchle, 1987; Cantox, 2002]. Through intact skin, 10–25% of the administered dose is absorbed [Black et al., 1975].

Ecotoxicity

The majority of bacteria are not able to metabolize triclosan [Vao and Salkinoja-Salonen, 1986; Liaw and Srinivasan, 1990; Schmidt et al., 1992, 1993]; therefore triclosan is detectable in aquatic ecosystems, sediments and sewage sludge [Tulp et al., 1979; Hites and Lopez-Avila, 1979; Lopez-Avila and Hites, 1980; Miyazaki et al., 1984; Paxéus, 1996]. Voets et al. [1976] described a 50% degradation of triclosan in an activated sewage model within 3 weeks. Further information regarding biological degradation was published by Hundt et al. [2000], Hay et al. [2001], Meade et al. [2001], McBain et al. [2003] and Schultz [2004]. Paxéus [2004] detected a degradation rate of >90% in biological sewage systems. Even under anaerobic conditions (sludge decomposition) a degradation of about 35% is possible [Räuchle, 1987].

Due to the high rate of photodegradation with a half-life of 3 h in summer [Räuchle, 1987], concentrations in surface water of 50 ng/l are substantially low [Singer et al., 2002].

Triclosan in sewage can exert negative effects on freshwater algae, nitrifying bacteria [Wilson et al., 2003; Dokianakis et al., 2004] and fish [Räuchle, 1987].

Concentrations up to 2 mg triclosan/l (limit of solubility) in sewage did not negatively influence the biological treatment of the sewage plants. Triclosan is highly toxic to fish (LC_{50} at 48 h dwell period 0.6 mg/l, threshold for harmfulness approx. 0.4 mg/l). In shorter dwell periods higher concentrations are tolerated, e.g. 10 mg/l at 5 min, 5 mg/l at 15 min, 2 mg/l at 30 min. The degradation products, which are induced under the influence of light, are less toxic to fish. In reality the concentrations appearing in sewage are at least 1 or 2 log steps below the LC_{50} [Räuchle, 1987]. Plants can accumulate triclosan [Nishina et al., 1991; von Woedtke et al., 1999].

The possibility of contamination of triclosan-containing products with dioxin represents a critical issue. Therefore any contamination with furans and dioxins has to be excluded, and a negative declaration by the manufacturer is required.

Other Biocides

Other relevant substances used for the impregnation of textiles are silver and copper compounds, chitosan, zeolites, quats, imidazoles, imidazolidinones, isothiazolines, curcumin, thujopsene, hinokitiol (from cypress oil), sulfadiazine, 1,3-dimethylol-5,5-dimethylhydantoin, thiobisphenol, neomycin and dimethyltetradecyl [3-(trimethoxysilyl)propyl]ammonium chloride and polyhexamethylene biguanide hydrochloride (polyhexanide) [Sun et al., 2001; Nakashima et al., 2002; Takai et al., 2002; Qian and Sun, 2003, 2004; Thiemann, 2004; Han and Yang, 2005; http://www.der-gruene-faden.de/ teyt/text2497.html, www.bgw.de]. As scientific data are difficult to obtain, the following characteristics (tables 4–6) are incomplete and underline the need for further scientific evaluation and documentation.

The problem of different activities obtained with various textiles is exemplarily described in the following experiments.

The textile was cut into small pieces (2 × 2 cm), autoclaved after washing with household laundry detergent, dried and incubated with test organisms. After exposure, viable bacterial cells were counted. Relative humidity was maintained at 100% for wet conditions and 50% for dry conditions at 22°C.

Under dry conditions Gram-negative rods rapidly lost their viability. Ag. Zn, ammonium zeolite and chitosan were proven to be effective against *S. aureus* for up to 6 h of incubation under wet and dry conditions, and also effective against MRSA for up to 24 h only under wet conditions (table 7). The results indicated significant differences between the tested textiles; therefore products should be tested by use of the same standard; however, a DIN EN ISO is presently not available.

Surprisingly silver and copper exhibit marked differences in tolerance although both are heavy metals with partly comparable chemical properties (table 5). In the aquatic toxicity these differences are not primarily remarkable (table 6).

Table 4. Spectrum of activity of selected biocides for textile impregnation

Agent	Bacteria[1]		Fungi[1]	Resistance
	Gram-positive	Gram-negative		
Triclosan	active	lack of activity	active	in vitro inducible
Silver	highly active	highly active	highly active	no data
Quats	active	lack of activity	lack of activity	in vitro inducible
Chitosan	active	active	active	no data
Polyhexanide	active	active	active	no data
Copper compounds (e.g. sulfide and sulfate)	active	active	active	positive [Aarestrup et al., 2002; Hasman and Aarestrup, 2002, 2005; Ugur and Ceylan, 2003]

[1]Studies on antimicrobial activity are not comparable due to differences in methodology.

Silver

The therapeutic activity of silver has been known since the 5th century. Ciro the Great ordered his troops to transport water in silver pots to protect the drinking water against spoilage and to preserve its potability.

In contrast to triclosan, nanocristalline silver is equally active against Gram-positive and Gram-negative bacteria, fungi and several viruses, and surpasses triclosan by antimicrobial activity [Thurman and Gerba, 1989; Ugur and Ceylan, 2003]. The concentration of 0.2 μg Ag/ml eradicates the amount of *E. coli* by 2 log steps within 13 min [Wuhrman and Zobrist, 1958]. After 3 h no bacterial growth was detectable on a polyurethane catheter contaminated with 10^9 CFU of *E. coli, P. aeruginosa, Enterobacter cloacae, Citrobacter freundii, C. albicans* or *Candida glabrata* [Guggenbichler, unpubl.].

Ag^+ is highly toxic to fish (table 6). The key mechanism of acute silver toxicity consists in the reduction of Na^+ uptake by blocking of Na^+, K^+-ATPase [Bianchini et al., 2002]. Like other metals, silver accumulates in aquatic food chains and may exert toxicity, which cannot be predicted from exposure to dissolved Ag^+ [Hook and Fisher, 2001].

For impregnation of textiles, silver can be anchored on fiber polymers so that the silver is not removable from the fibers. As long as no silver is released, no toxic risks are given.

Quaternary Ammonium Compounds

The attachment of a quat to silk fibers could be realized by its polymerization. This new tissue combines the favorable properties of silk with protection

Table 5. Tolerance to selected textile biocides

Agent	Dermal resorption	Toxicity	Allergenicity	Mutagenicity	Carcinogenicity	Teratogenicity
Triclosan	yes	little to nontoxic	low	neg.	neg.	neg.
Silver	no	depends on liberation of compound: little to nontoxic	no	neg.	neg.	neg.
Quats	yes	moderate to highly toxic	moderate	neg.	neg.	neg.
Chitosan	no data	no hint (biologically deduced)	no	neg.	no data	no data
Polyhexanide	no	little to nontoxic	no	neg.	neg.	no teratogenic hazard to humans
Copper compounds (e.g. sulfide and sulfate)	no	essential trace element, dose-dependent toxicity	no	partly neg., but potentially mutagenic [Agarwal et al., 1989]	not cocarcinogenic [Sunderman et al., 1974]	neg. (2 mg Cu/kg) to rats [Mason et al., 1989] and humans [Kahn-Nathan, 1975; Barash et al., 1990], pos. to crab and copepods at environmental concentrations [Fisher and Hook, 2002; Lavolpe et al., 2004], shrimp [Rayburn and Aladdin, 2003] and *Chironomus tentans* [Martinez et al., 2003]

Table 6. Ecotoxicity of selected textile biocides

Agent	Biodegradation	Aquatic toxicity
Triclosan	slow	high toxicity for fish, modified behavior and growth of *Rana pipiens* in OWC concentrations [Fraker and Smith, 2004], 48-hour EC_{50} to *Daphnia magna* 340 μ/l, 96-hour LC_{50} to *Pimephales promelas* and *Lepomis macrochirus* 260 and 370 μ/l, respectively, NOEC and LOEC to *Oncorhynchus mykiss* 34.1 and 71.3 μ/l [Orvos et al., 2002]; 96-hour LC_{50} for 24-hour-old larvae of *Oryzias latipes* 602 μg/l, increased concentration of hepatic vitellogenin (estrogenic effect) at 20 μg/l with adverse effects in F_1 generation [Ishibashi et al., 2004]
Silver	no	$AgNO_3$ 48-hour LC_{50} to *Ceriodaphnia dubai* 0.5 μ/l, 8-day LC_{50} 0.32 μ/l, NOEC of silver cysteinate <0.001 μg/l, LOEC of $AgNO_3$ and silver glutathionate 0.01 and 0.6 μ/l, respectively [Bielmyer et al., 2002], 96-hour LC_{50} 330–2,700 μg/l to trout *Oncorhynchus mykiss* in seawater and 5–70 μg/l in freshwater [Hogstrand and Wood, 1998]
Quats	depends on structure: slow to rapid	dependent on structure: highly toxic
Chitosan	good	sublethal concentration to carp 75–150 mg/l [Dautremepuits et al., 2004]
Polyhexanide	no	highly toxic to fish
Copper compounds	no	$CuCl_2$ *Lemna minor* EC_{50} 0.3–0.9 mg/l, 1 ppm sublethal to algae [Mal et al., 2002], sublethal concentration to carp 0.1 mg/l [Dautremepuits et al., 2004], LOEC in water 6.8–13.6 μg/l [Hsieh et al., 2004]

LOEC = Lowest observed effect concentration; NOEC = no observed effect concentration; OWC = organic wastewater contaminant.

from microbial colonization. At longer contact times nonpolymerized quats can be absorbed by skin [BIA, 1995]. On the basis of their LD_{50} (and their chemical structure) and depending on the application mode, quats can be qualified as mild to highly toxic [Kramer et al., 1985; Merck Schuchardt, 2001]. In vitro benzalkonium chloride, a frequently used quat, irreversibly damages ciliated nasal epithelia [Klöcker and Rudolph, 2000]. Also wound healing is retarded [Bolton et al., 1985]. Quats have a potential for sensitization. Benzalkonium chloride is highly toxic for aquatic organisms [Material Safety Data Sheet 1999], but the agent is degraded within 28 days [Zöllner et al., 1995]. The development of antibacterial resistance is possible [Rudolf and Kampf, 2003].

Resuming these data a prolonged dermatological evaluation of benzalkonium chloride is not justified [Kramer et al., 2003]. This is also valable for other non-polymerized quats. These findings cannot generally be transferred to polymerized quats but require adequate investigations of the long-term tolerance of humans as

Table 7. Antibacterial properties of antimicrobial-impregnated textiles [Takai et al., 2002]

Antimicrobial agent	Efficacy				
	S. aureus		MRSA		*P. aeruginosa*
	dry	wet	dry	wet	wet
Ag, Zn, Cu zeolite	++	++	–	++	+
Ag, Zn, ammonium zeolite	+++	+++[1]	–	++++[1]	++[1]
Aliphatic imide	–	–	–	+	+
Quat	–	–	–	+	+
Chitosan	+++	+++[2]	–	++[2]	++[1]
Without	–	–	–	–	–

– = Not effective; + = slightly effective; ++ = moderately effective; +++ = effective; ++++ = highly effective.

[1]Significantly decreased activity with organic load.

[2]No influence of organic load.

well as of the environment. Beside this, the stability against microbial degradation of the polymer in textiles during wearing and washing must be analyzed.

The use of quats in the treatment of atopic dermatitis reduced the SCORAD within 1 week from 43 to 30 ($p = 0.003$). The local score improved from 32 to 18.6 ($p = 0.001$), the controls remained unchanged [Ricci et al., 2004]. Further clinical trials and the comparison with other therapeutic options are warranted.

Due to the risk of resistance development, any attempts to combine e.g. cephalosporins with quats are not indicated [Kim et al., 2000].

Chitosan

Chitosan is nowadays used in wound dressings and textiles [Allan et al., 1984; Li et al., 1992; Hirano, 1996; Tokura et al., 1996; Lee et al., 1997; Nam et al., 2001].

Chitosan is the deacetylated compound of chitin [Lim and Hudson, 2004] and as natural biopolymer can be extracted from the shells of aquatic animals (crabs, shrimp shells). The amino group in the C2 position of the cationic glucosamine provides its antimicrobial activity [Chen et al., 2002; Al-Bahra, 2004] by binding to the bacterial cells [Knobelsdorf and Mieck, 2000; Takai et al., 2002; Al-Bahra, 2004].

In a quantitative suspension test, chitosan is highly active. In 10 min a 0.1% solution achieves a reduction of *S. aureus* and *P. aeruginosa* of ≥5 log, of

C. albicans one of ≥4 log [Weber, unpubl.]. It is noteworthy that in the test chitosan remained undissolved as suspension, a situation comparable to textiles.

A decrease in antimicrobial activity by up to 2 log steps is observed in the presence of hyaluronic acid, ascorbic acid and NaCl, which has to be taken into consideration for practical purposes.

No allergic reactions are reported at present [Knittel and Schollmeyer, 1998]. On the basis of the LD_{50} in the mouse of 16 g/kg BW orally, chitosan can be considered as nontoxic. Adverse effects were observed at concentrations of >2,000 mg/kg BW in rats [Kim et al., 2001]. In accordance with the results of toxicological investigations, wound dressings with chitosan and collagen were proven to be favorable [Ye et al., 2004]. Because the agent is used for food conservation and weight reduction (weight loss supplements), no toxicity for impregnated textiles can be expected. In vitro tests with tissue cultures (hepatocytes) assumed this point of view [Risbud et al., 2003]. Furthermore, chitosan is antigenotoxically active by adsorbing mutagens [Ohe, 1996].

Chitosan is biodegradable and bioabsorbable [Pascual and Julia, 2001; Al-Bahra, 2004] and can be expected to play a role in the antioxidative mechanism of biological systems [Xue et al., 1998].

Zeolites

Zeolites are composed of silicium and aluminium tetrahedrons. By oxygen bridges these form secondary complexes agglomerating to tertiary tetrahedrons. Thereby a crystal structure with an extensive system of pores and channels is formed. As ion exchangers zeolites are endowed with antimicrobial agents, e.g. silver for impregnation of catheters and textiles [Thom et al., 2003].

Zeolites are not water soluble. The basic structure is nontoxic. In sewage sludge zeolites are degraded to silicic acid (http://sodasan.com/diplomarbeit/Pla-compaktp4.htm'#T3). Toxic and ecotoxic properties of zeolites originate exclusively from the coupled biocides.

Polyhexanide

The cationic and polymeric structure of the agent allows it to bind strongly to cellulosic fabrics such as cotton and viscose [Payne, 1997]. The good activity against both Gram-positive and Gram-negative bacteria has led to its use on cotton wound dressings and drain sponges. Cotton gauze containing polyhexanide has been found to be an effective barrier to bacterial penetration, even in the presence of protein [Reitsma and Rodeheaver, 2001]. A dressing substantially reduced (by 4 or 5 log) the amount of *P. aeruginosa* that had gained access to the wound bed and reduced the bacterial inoculum in the dressing itself. The presence of polyhexanide did not prevent epithelialization of partial-thickness wounds.

A repeated-insult patch test effectively demonstrated that the polyhexanide-treated gauze is devoid of skin-sensitizing properties [Orr et al., 2001].

Copper

On the basis of its broad antibacterial and antifungal activity, this agent was used in surgical gloves, filters, socks and antimite mattresses. In animal experiments no potential for sensibilization was detected [Borkow and Gabbay, 2004]. While copper is nonmutagenic for mammals, the compound is teratogenic and embryotoxic for aquatic organisms (table 6). For the estimation of ecotoxicity the evidence of available investigations in mammals is not sufficient. With the use in filters as exception, the clinical application of this agent is questionable.

Obsolete Compounds

The following agents, which have been described by Wigert et al. [1974] for use as antimicrobial impregnation of household and technical textiles, are considered obsolete: hexachlorophene, chlorphenols, organomercury, -tin and -zinc compounds, anorganic silver salts, hydroxychinolines, salicylanilides, dithiocarbamates, tetramethylthiuramdisulfide, neomycin and tetracycline.

In Germany *tributyl tin compounds* (TBT) are not allowed for antimicrobial impregnation of textiles. Otherwise, for heavy textile tissues, e.g. tents, its use is allowed. Due to its environmental burden this use has also to be dismissed because of its estrogen-like activity.

The concentrations of TBT in textiles ($<$110 mg/kg) are below the microbicidal concentration of 0.1% and derive from stabilizers and catalysts. Their maximal concentrations, which are reached by dermal absorption, are clearly below the tolerable daily intake value recommended by the WHO (www.bgvv.de, TBT and other organic tin compounds in foodstuff and consumer products, March 6, 2000). This evaluation solely takes the isolated action of the compound on the organism into consideration, whereas in reality human exposure always results from a mixture of compounds with unknown interactions. Because a toxicological analysis of various combinations with TBT has not been performed, a final risk assessment is not possible. For safety reasons, TBT generally has to be abandoned also as auxiliary material for textiles.

Textiles, manufactured in Germany, can be assumed as free of insecticides, pesticides and fungicides. In imported textiles these agents cannot be excluded because of the application for protection of moulding and bacterial destruction, especially as impregnations in tropical climates. Especially in imported textiles from developing countries, antimicrobial compounds, restricted in the EU, can be found, since only polychlorated phenols (e.g. pentachlorphenol) are forbidden for import materials. To remove unwanted residues, it is generally recommended to wash, to clean and to air imported textiles before wearing (http:// www.swisstextiles.ch/boxalino/files/Document183file.pdf).

Odor-Absorbing Agents

Cyclodextrin

Cyclodextrins are natural hydrophilic complex-building agents, which absorb odors without antimicrobial activity. Since 2000 these compounds have been allowed as food additives [Buschmann et al., 2001, 2003; Knittel et al., 2004]. As long as these compounds are not combined with biocides (e.g. with isothiazolinone), they can be used without risks for humans and the environment.

Conclusion

For toxicological reasons for humans and the environment the addition of so-called antismell agents with antimicrobial properties to textiles for daily use has to be avoided. Smelling textiles have to be returned to hygienic conditions, e.g. by washing. Textiles impregnated with antismell agents tend to be cleaned less frequently with the result of loss of functional quality. For the intended inhibition of malodor, deodorants or antiperspirants are the agents of choice.

Insofar as therapeutic or prophylactic advantages of antimicrobial textiles are proven and this application is at least equally effective and tolerated compared to conventional therapeutic options, no objections against their use can be made. In any case the used biocide has to be declared with documented efficacy. As these products represent almost medical devices or medicinal products in case of addition of a drug to textiles in the near future, the conformity of the registration needs legal assessment by notified bodies or by the drug approval authorities [Kramer et al., 2005a, b].

At present, important deficits are to be mentioned:

- International norms for in vitro testing as well as for preclinical assessment (antimicrobial activity and tolerance) of antimicrobial-impregnated textiles are missing. Compared with the harmonized European norms for testing of disinfectants, this situation warrants an urgent amendment.
- As for medical devices, independently from the intended use, a declaration (agent and concentration) for antimicrobial-impregnated textiles is requested. To prevent inappropriate use, a registration also for consumer products is wanted.
- For every agent a documented risk-benefit assessment has to be requested, including the intended application as well as the long-term tolerance by humans and the environment.

References

Aarestrup FM, Hasman H, Jensen LB, Moreno M, Herrero IA, Dominguez L, Finn M, Franklin A: Antimicrobial resistance among enterococci from pigs in three European countries. Appl Environ Microbiol 2002;68:4127–4129.

Agarwal K, Sharma A, Talukder G: Effects of copper on mammalian cell components. Chem Biol Interact 1989;69:1–16.

Al-Bahra MM: Darstellung von Chitinderivaten zur antimikrobiellen Ausrüstung von Textilien; dissertation, Rheinisch-Westfälische Technische Hochschule, Aachen, 2004.

Allan GG, Altman LC, Bensinger RE, Ghosh DK, Hirabayashi Y, Neogi AN, Neogi S: Fragmentation of chitosan by microfluidization process; in Zikakis JP (ed): Chitin, Chitosan and Related Enzymes. Orlando, Academic Press, 1984, pp 173–185.

Arnold LB Jr: Antimicrobial finishes for textiles: some practical aspects. Am Dyestuff Rep 1963;52:192–195.

Bager P, Westergaard T, Rostgaard K, Hjalgrim H, Melbye M: Age at childhood infections and risk of atopy. Thorax 2002;57:379–382.

Barash A, Shoham Z, Borenstein R, Nebel L: Development of human embryos in the presence of a copper intrauterine device. Gynecol Obstet Invest 1990;29:203–206.

Barbolt TA: Chemistry and safety of triclosan, and its use as an antimicrobial coating on coated Vicryl Plus antibacterial suture (coated polyglactin 910 suture with triclosan). Surg Infect 2002;3(suppl 1):45–54.

BIA (Berufsgenossenschaftliches Institut für Arbeitssicherheit): GESTIS – Stoffdatenbank: Alkylbenzyldimethylammoniumchlorid. Sankt Augustin, 1995.

Bianchini A, Grosell M, Gregory SM, Wood CM: Acute silver toxicity in aquatic animals is a function of sodium uptake rate. Environ Sci Technol 2002;36:1763–1766.

Bielmyer GK, Bell RA, Klaine SJ: Effects of ligand-bound silver on *Ceriodaphnia dubia.* Environ Toxicol Chem 2002;21:2204–2208.

Black JG, Howes D, Rutherford T: Percutaneous absorption and metabolism of Irgasan DP300. Toxicology 1975;3:33–47.

Bodner C, Godden D, Seaton A: Family size, childhood infections and atopic diseases. The Aberdeen WHEASE Group. Thorax 1998;53:28–32.

Bolton L, Oleniacz W, Constantine B, Kelliher BO, Jensen D, Means B, Rovee D: Repair and antibacterial effects of topical antiseptic agents in vivo; in Maibach HI, Lowe NJ (eds): Models in Dermatology. Basel, Karger, 1985, vol 2, pp 145–158.

Borkow G, Gabbay J: Putting copper into action: copper-impregnated products with potent biocidal activities. FASEB J 2004;18:1728–1730.

Braid JJ, Wale MC: The antibacterial activity of triclosan-impregnated storage boxes against *Staphylococcus aureus, Escherichia coli, Pseudomonas aeruginosa, Bacillus cereus* and *Shewanella putrefaciens* in conditions simulating domestic use. J Antimicrob Chemother 2002;49:87–94.

Braoudaki M, Hilton AC: Adaptive resistance to biocides in *Salmonella enterica* and *Escherichia coli* O157 and cross-resistance to antimicrobial agents. J Clin Microbiol 2004;42:73–78.

Braoudaki M, Hilton AC: Mechanisms of resistance in *Salmonella enterica* adapted to erythromycin, benzalkonium chloride and triclosan. Int J Antimicrob Agents 2005;25:31–37.

Brenwald NP, Fraise AP: Triclosan resistance in methicillin-resistant *Staphylococcus aureus* (MRSA). J Hosp Infect 2003;55:141–144.

Buschmann HJ, Knittel D, Schollmeyer E: Textile Materialien mit Cyclodextrinen. Melliand Textilber 2001;5:368–370.

Buschmann HJ, Knittel D, Schollmeyer E: Wie funktionieren Textilien mit Cyclodextrinen? Melliand Textilber 2003;84:988–989.

Cantox US Inc: Absorption, distribution, metabolism, and excretion profile of triclosan and risk evaluation of triclosan for use in absorbable sutures, 2002.

Chambers PR: FAT 80'023/S potential tumorigenic and chronic toxicity effects in prolonged dietary administration to hamsters. Ciba Drug Master File, 1999.

Chen YM, Chung YC, Wang LW, Chen KT, Li SY: Antibacterial properties of chitosan in waterborne pathogen. J Environ Sci Health Part A Tox Hazard Subst Environ Eng 2002;37:71379–1390.

Chow AY, Hirsch GH, Buttar HS: Nephrotoxic and hepatotoxic effects of triclosan and chlorhexidine in rats. Toxicol Appl Pharmacol 1977;42:1–10.

Chuanchuen R, Beinlich K, Hoang TT, Becher A, Karkhoff-Schweizer RR, Schweizer HP: Cross-resistance between triclosan and antibiotics in *P. aeruginosa* is mediated by multidrug efflux pumps: exposure of a susceptible mutant strain to triclosan selects nfxB mutants overexpressing MexCD-OprJ. Antimicrob Agents Chemother 2001;45:428–432.

Chuanchuen R, Karkhoff-Schweizer RR, Schweizer HP: High-level triclosan resistance in *Pseudomonas aeruginosa* is solely a result of efflux. Am J Infect Control 2003;31:124–127.

Chuanchuen R, Narasaki CT, Schweizer HP: The MexJK efflux pump of *Pseudomonas aeruginosa* requires OprM for antibiotic efflux but not for efflux of triclosan. J Bacteriol 2002;184: 5036–5044.

Cookson BD, Farrelly H, Stapleton P, Garvey RPJ, Price MR: Transferable resistance to triclosan in MRSA. Lancet 1991;337:1548–1549.

Cookson OC, Moffatt MF: Asthma: an epidemic in the absence of infections? Science 1997;275:41–42.

Dautremepuits C, Betoulle S, Paris-Palacios S, Vernet G: Immunology-related perturbations induced by copper and chitosan in carp (*Cyprinus carpio* L.). Arch Environ Contam Toxicol 2004;47: 370–378.

Denning HJ, Sliwa S, Wilson GA: Triclosan: effects on pregnancy and post-natal development in rats. Ciba Drug Master File, 1992.

De Salva SJ, Kong BM, Lin YJ: Triclosan: a safety profile. Am J Dent 1989;2(spec iss):185–196.

De Vizio W, Davies R: Rationale for the daily use of a dentifrice containing triclosan in the maintenance of oral health. Compend Contin Educ Dent 2004;25(suppl 1):54–57.

Dokianakis SN, Kornaros ME, Lyberatos G: On the effect of pharmaceuticals on bacterial nitrite oxidation. Water Sci Technol 2004;50:341–346.

Drake JC: 1-year oral toxicity study in baboons with compound FAT 80'023/A. Ciba Drug Master File, 1975.

Elek SD, Conen PE: The virulence of *Staphylococcus pyogenes* for man: a study of problems with wound infection. Br J Exp Pathol 1957;38:573–586.

Engel RA, Gump WS: Antiseptics for textile purposes. Am Dyestuff Rep 1941;30:163–165.

Farooqi IS, Hopkin JM: Early childhood infection and atopic disorder. Thorax 1998;53:927–932.

Fisher NS, Hook SE: Toxicology tests with aquatic animals need to consider the trophic transfer of metals. Toxicology 2002;181–182:531–536.

Fraker SL, Smith GR: Direct and interactive effects of ecologically relevant concentrations of organic wastewater contaminants on *Rana pipiens* tadpoles. Environ Toxicol 2004;19:250–256.

Friedrich N, Völzke H, Schwahn C, Kramer A, Jünger M, Schäfer T, John U, Kocher T: Inverse association between periodontitis and respiratory allergies. Clin Exp Allergy 2006;36:495–502.

Furuichi Y, Rosling B, Volpe AR, Lindhe J: The effect of a triclosan/copolymer dentifrice on healing after non-surgical treatment of recurrent periodontitis. J Clin Peridontol 1999;26:63–66.

Gauger M, Mempel M, Schekatz A, Schäfer TH, Ring J, Abeck D: Silberbeschichtete Textilien reduzieren die *Staphylococcus-aureus*-Besiedelung bei Patienten mit atopischem Ekzem. Dermatology 2003;207:15–21.

Gibbs S, Surridge H, Adamson R, Cohen B, Bentham G, Reading R: Atopic dermatitis and the hygiene hypothesis: a case-control study. Int J Epidemiol 2004;33:199–207.

Gilbert P, McBain A: Literature-based evaluation of the potential risks associated with impregnation of medical devices and implants with triclosan. J Surg Infect 2002;3(suppl 1):S55–S63.

Gloor M, Becker A, Wasik B, Kniehl E: Triclosan, ein dermatologisches Lokaltherapeutikum. Hautarzt 2002;53:724–729.

Gocke E, King MT, Eckhardt K, Wild D: Mutagenicity of cosmetic ingredients licensed by the European Communities. Mutat Res 1981;90:91–109.

Goldsmith L: 90-day oral toxicity study in rats with FAT 801023H. Ciba Drug Master File, 1983.

Goodfellow G, Lee-Brotherton V, Daniels J, Roberts A, Nestmann E: Antibacterial resistance and triclosan. Soc Toxicol Ann Meet, Salt Lake City, 2003.

Goodman DG: Pathology Working Group report on triclosan chronic toxicity/carcinogenicity study in Sprague-Dawley rats. Ciba Drug Master File, 1990.

Han SY, Yang Y: Antimicrobial activity of wool fabric treated with curcumin. Dyes Pigment 2005;64: 157–161.

Hanioka N, Omae E, Nishimura T, Jinno H, Onodera S, Yoda R, Ando M: Interaction of 2,4,4′-trichloro-2′-hydroxydiphenyl ether with microsomal cytochrome P450-dependent monooxygenases in rat liver. Chemosphere 1996;3:265–276.

Hasman H, Aarestrup FM: *tcrB*, a gene conferring transferable copper resistance in *Enterococcus faecium*: occurrence, transferability, and linkage to macrolide and glycopeptide resistance. Antimicrob Agents Chemother 2002;46:1410–1416.

Hasman H, Aarestrup FM: Relationship between copper, glycopeptide, and macrolide resistance among *Enterococcus faecium* strains isolated from pigs in Denmark between 1997 and 2003. Antimicrob Agents Chemother 2005;49:454–456.

Hay AG, Dees PM, Sayler GS: Growth of a bacterial consortium with triclosan. FEMS Microbiol Ecol 2001;36:105–112.

Heath RJ, Rock CO: A triclosan-resistant bacterial enzyme. Nature 2000;406:145–146.

Henderson LM, Produlock RJ, Haynes P: Mouse micronucleus test on triclosan. Ciba Drug Master File, 1988a.

Henderson LM, Ransome SJ, Brabbs CE: An assessment of the mutagenic potential of triclosan using the mouse lymphoma TK locus assay. Ciba Drug Master File, 1988b.

Hidalgo E, Bartolome R, Barroso C, Moreno A, Dominguez C: Silver nitrate: antimicrobial activity related to cytotoxicity in cultured human fibroblasts. Skin Pharmacol Appl Skin Physiol 1998;11:140–151.

Hindson TC: Irgasan DP 300 in a deodorant. Contact Dermatitis 1975;1:328.

Hirano S: Chitin biotechnology applications. Biotechnol Annu Rev 1996;2:237–258.

Hites RF, Lopez-Avila V: Identification of organic compounds in an industrial wastewater. Anal Chem 1979;51:1452A–1456A.

Hogstrand C, Wood CM: Toward a better understanding of the bioavailability, physiology and toxicity of silver in fish: implications for water quality criteria. Environ Toxicol Chem 1998;17:547–561.

Hook SE, Fisher NS: Sublethal effects of silver in zooplankton: importance of exposure pathways and implications for toxicity testing. Environ Toxicol Chem 2001;20:568–574.

Hsieh CY, Tsai MH, Ryan DK, Pancorbo OC: Toxicity of the 13 priority pollutant metals to *Vibrio fisheri* in the Microtox chronic toxicity test. Sci Total Environ 2004;320:37–50.

Hundt K: Biotransformation von halogenierten Diphenylethern durch Pilze unter besonderer Berücksichtigung von *Trametes versicolor*; dissertation, Ernst-Moritz Arndt University, Greifswald, Math-Nat Fac, 2001.

Hundt K, Martin D, Hammer E, Jonas U, Kindermann MK, Schauer F: Transformation of triclosan by *Trametes versicolor* and *Pycnoporus cinnabarinus*. Appl Environ Microbiol 2000;66:4157–4160.

Innes ME, Umraw N, Fish JS, Gomez M, Cartotto RC: The use of silver coated dressings on donor site wounds: a prospective, controlled matched pair study. Burns 2001;27:621–627.

Ishibashi H, Matsumura N, Hirano M, Matsuoka M, Shiratsuchi H, Ishibashi Y, Takao Y, Arizono K: Effects of triclosan on the early life stages and reproduction of medaka *Oryzias latipes* and induction of hepatic vitellogenin. Aquatic Toxicol 2004;67:167–179.

Jones E, Wilson L: Ames metabolic activation test to address the potential mutagenic effect of triclosan. Ciba Drug Master File, 1988.

Jünger M, Daeschlein G, Ladwig A, Nauck M, Haase H, Kramer A: Reduction of staphylococci in neurodermitic lesions after wearing of silver-impregnated textile. In preparation.

Kahn-Nathan J: What evidence is there to judge the effect of copper in the body? Contracept Fertil Sex (Paris) 1975;3:346.

Kim SK, Park PJ, Yang HP, Han SS: Subacute toxicity of chitosan oligosaccharide in Sprague-Dawley rats. Arzneimittelforschung 2001;51:769–774.

Kim SH, Son H, Nam G, Chi DY, Kim JH: Synthesis and in vitro antibacterial activity of 3-[N-methyl-N-(3-methyl-1,3-thiazolium-2-yl)amino]methyl cephalosporin derivatives. Bioorg Med Chem Lett 2000;10:1143–1145.

Kligman AM, Breit R: The identification of phototoxic drugs by human assay. J Invest Dermatol 1968;51:90–99.

Klöcker N, Rudolph P: Konservierte Nasensprays sind obsolet. Pharm Z 2000;145:40–42.

Knittel D, Buschmann HJ, Hipler C, Elsner P, Schollmeyer E: Funktionelle Textilien zur Hautpflege und als Therapeutikum. Akt Dermatol 2004;30:11–17.

Knittel D, Schollmeyer E: Chitosan und seine Derivate für die Textilveredlung. 1. Ausgangssituation. Textilveredlung 1998;33:67–71.

Knobelsdorf C, Mieck KP: Hygienisch wirkende Garne durch den Einsatz von Chitosanfasern. Textilveredlung 2000;7/8:10–15.
Kramer A: Antiseptika und Händedesinfektionsmittel; in Korting HC, Sterry W (eds): Therapeutische Verfahren in der Dermatologie – Dermatika und Kosmetika. Berlin, Blackwell Wissenschaft, 2001, pp 273–294.
Kramer A, Assadian O, Guggenbichler P, Heidecke CD, Jünger M, Lippert H, Schauer F: Beurteilung von Triclosan bezüglich seines Einsatzes in chirurgischem Nahtmaterial und Konsequenzen aus den Wirkstoffeigenschaften für den medizinischen und nicht medizinischen Einsatz, 2005a. www.dgkh.de
Kramer A, Assadian O, Schneider I, Soltau U, Pitten FA, Schwemmer J, Siegmund W, Müller G, Jäkel C: Begründung zur Einordnung Polihexanid-haltiger Wundspüllösungen als Arzneimittel. Stellungnahme zum Beitrag von K. Kaehn, Zur Einstufung von Polyhexanid-haltigen Wundspüllösungen, ZfW 2005;3:124–127. ZfW 2005b;4:185–187.
Kramer A, Berencsi G, Weuffen W: Toxische und allergische Nebenwirkungen von Antiseptika; in Weuffen W, Berencsi G, Kramer A, Gröschel D, Kemter BP, Krasilnikow AP (eds): Handbuch der Antiseptik. Stuttgart, Fischer, 1985, vol I/5, pp 113–210.
Kramer A, Daeschlein G, Kammerlander G, Andriessen A, Aspöck C, Bergemann R, Eberlein T, Gerngross H, Görtz G, Heeg P, Jünger M, Koch S, König B, Laun R, Peter RU, Roth B, Ruef C, Sellmer W, Wewalka G, Eisenbeiss W: Konsensusempfehlung zur Auswahl von Wirkstoffen für die Wundantiseptik. Hyg Med 2004;29:147–157.
Kramer A, Mersch-Sundermann, Gerdes H, Pitten F-A, Tronnier H: Toxikologische Bewertung für die Händedesinfektion relevanter antimikrobieller Wirkstoffe; in Kampf G (eds): Hände-Hygiene im Gesundheitswesen. Berlin, Springer, 2003, pp 105–160.
Kraus H, Tiefenbrunner F: Stichprobenartige Überprüfung einzelner Tiroler Schwimmbäder auf das Vorkommen von *Trichomonas vaginalis* und von pathogenen Pilzen. Zentralbl Bakt Hyg I Abt Orig B 1975;160:286–291.
Lachapelle JM, Tennstedt D: Low allergenicity of triclosan: predictive testing in guinea pigs and in humans. Dermatologica 1979;158:379–383.
Lavolpe M, Greco LL, Kesselman D, Rodriguez E: Differential toxicity of copper, zinc, and lead during the embryonic development of *Chasmagnathus granulatus* (Brachyura, Varunidae). Environ Toxicol Chem 2004;23:960–967.
Lee JW, Nam CW, Ko SW: Antimicrobial finish of cotton fabric with chitosan oligomer. Proc 4th Asian Textile Conf, Federation of Asian Professional Textile Association, Taipei, June 1997, vol 2, pp 845–848.
Levy CE, Roujeinikova A, Sedelnikova S, Baker PJ, Stuitje AR, Slabas AR, Rice DW, Rafferty JB: Molecular basis of triclosan activity. Nature 1999;398:383–384.
Levy SB: Antibacterial household products: cause for concern. Emerg Infect Dis 2001;7:512–515.
Li Q, Grandmaison EW, Goosen MFA, Dunn ET: Applications and properties of chitosan. J Bioact Compat Polym 1992;7:370–397.
Liaw HJ, Srinivasan VR: Expression of an *Erwinia* sp. gene encoding diphenyl ether cleavage in *Escherichia coli* and an isolated *Acinetobacter* strain PE7. Appl Environ Microbiol 1990;32: 686–689.
Lim SH, Hudson SM: Synthesis and antimicrobial activity of a water-soluble chitosan derivative with a fiber-reactive group. Carbohydr Res 2004;339:313–319.
Lopez-Avila V, Hites RA: Organic compounds in an industrial wastewater: their transport into sediments. Environ Sci Technol 1980;14:1382–1390.
McBain AJ, Bartolo RG, Catrenich CE, Charbonneau D, Ledder RG, Price BB, Gilbert P: Exposure of sink drain microcosms to triclosan: population dynamics and antimicrobial susceptibility. Appl Environ Microbiol 2003;69:5433–5442.
McMurry LM, McDermott PF, Levy SB: Genetic evidence that InhA of *Mycobacterium smegmatis* is a target for triclosan. Antimicrob Agents Chemother 1999;43:711–713.
McMurry LM, Oethinger M, Levy SB: Triclosan targets lipid synthesis. Nature 1998;394:531–532.
Maitre S, Jaber K, Perrot JL, Guy C, Cambazard F: Increased serum and urinary levels of silver during treatment with topical silver sulfadiazine. Ann Dermatol Vénéréol 2002;129:217–219.
Mal TK, Adorjan P, Corbett AL: Effect of copper on growth of an aquatic macrophyte, *Elodea canadensis*. Environ Pollut 2002;120:307–311.

Marrakchi H, Zhang YM, Rock CO: Mechanistic diversity and regulation of type II fatty acid synthesis. Biochem Soc Trans 2002;30:1050–1055.
Martinez EA, Moore BC, Schaumloffel J, Dasgupta N: Morphological abnormalities in *Chironomus tentans* exposed to cadmium- and copper-spiked sediments. Ecotoxicol Environ Saf 2003;55:204–212.
Marzulli FN, Maibach HI: Antimicrobials: experimental contact sensitization in man. J Soc Cosmet Chem 1973;24:399–421.
Mason RW, Edwards IR, Fisher LC: Teratogenicity of combinations of sodium dichromate, sodium arsenate and copper sulphate in the rat. Comp Biochem Physiol C 1989;93:407–411.
Material Safety Data Sheet (MSDS): n-Propyl alcohol. Mallinckrodt Baker, 11/17/99. www.JTBaker.com
Matricardi PM: Infections preventing atopy: facts and new questions. Allergy 1997;52:879–882.
Matricardi PM, Bonini S: High microbial turnover rate preventing atopy: a solution to inconsistencies impinging on the hygiene hypothesis? Clin Exp Allergy 2000a;30:1506–1510.
Matricardi PM, Rosmini F, Riondino S, Fortini M, Ferrigno L, Rapicetta M, Bonini S: Exposure to foodborne and orofecal microbes versus airborne viruses in relation to atopy and allergic asthma: epidemiological study. BMJ 2000b;320:412–417.
Meade MJ, Waddell RL, Callahan TM: Soil bacteria *Pseudomonas putida* and Alcaligenes xylosoxidans subsp. *denitrificans* inactivate triclosan in liquid and solid substrates. FEMS Microbiol Lett 2001;204:45–48.
Merck Schuchardt: Sicherheitsdatenblatt Alkylbenzyldimethylammoniumchlorid, Ethanol, Ethylenglycolmonophenylether, 1-Propanol, 2-Propanol. February 2001. CD-ROM 2001/1D.
Meyer-Rohn J, Kulenkamp D: Klinik und Therapie der Dermatomykosen. Ärztl Praxis 1975;27: 108–110.
Mezcua M, Gomez MJ, Ferrer I, Aguera A, Hernando MD, Fernandez-Alba AR: Evidence of 2,7/2,8-dibenzodichloro-*p*-dioxin as a photodegradation product of triclosan in water and wastewater samples. Anal Chim Acta 2004;524:241–247.
Miyazaki T, Yamagashi T, Matsumoto M: Residues of 4-chloro-1-(2,4-dichlorophenoxy)-2-methoxybenzenes (triclosan methyl) in aquatic biota. Bull Environ Contam Toxicol 1984;32:227–232.
Morseth SL: Two-generation reproduction study in rats – FAT 80'023. Ciba Drug Master File, 1988.
Mosmann TR, Coffman RL: TH1 and TH2 cells: different patterns of lymphokine secretion lead to different functional properties. Annu Rev Immunol 1989;7:145–173.
Müller G, Kramer A, Karkour Y: Stellenwert silberbasierter Wundantiseptika. Vasomed, in press.
Müller G, Winkler Y, Kramer A: Antibacterial activity and endotoxin-binding capacity of Actisorb® Silver 220. J Hosp Infect 2003;53:211–214.
Nakashima H, Miyano N, Sawabe Y, Takatuka T: Photolysis and antimicrobial activity of hinokitiol in antimicrobial/deodorant processed textiles. Sen-I Gakkaishi 2002;58:129–134.
Nam CW, Kim YH, Ko SW: Blend fibers of polyacrylonitrile and water-soluble chitosan derivative prepared from sodium thiocyanate solution. J Appl Polymer Sci 2001;82:1620–1629.
Nishina A, Kihara H, Uchibori T, Oi T: Antimicrobial substances in 'DF-100' extracts of grapefruit seeds. Bokin Bobai 1991;19:401–404.
Ohe T: Antigenotoxic activities of chitin and chitosan as assayed by sister chromatid exchange. Sci Total Environ 1996;181:1–5.
Orr R, Eggleston T, Shelanski M: Determination of the irritating and sensitizing properties of Kerlix® AMD antimicrobial gauze dressing on scarified human skin, 2001. http://www.kendallhq.com/catalog/ClinicalInformation/KERLIXAMDonSkin.pdf
Orvos DR, Versteeg DJ, Inauen J, Capdevielle M, Rothenstein A, Cunningham V: Aquatic toxicity of triclosan. Environ Toxicol Chem 2002;21:1338–1349.
Pascual E, Julia MR: The role of chitosan in wool finishing. J Biotechnol 2001;89:289–296.
Paterson RA: 13-week oral toxicity study in rabbits. Ciba Drug Master File, 1969.
Paxéus N: Organic pollutants in effluents of large wastewater treatment plants in Sweden. Water Res 1996;30:1115–1122.
Paxéus N: Removal of selected non-steroidal anti-inflammatory drugs (NSAIDs), gemfibrozil, carbamazepine, beta-blockers, trimethoprim and triclosan in conventional wastewater treatment plants in five EU countries and their discharge to the aquatic environment. Water Sci Technol 2004;50:253–260.
Payne J: From medical textile to smell-free socks. J Soc Dyers Colour 1997;113:48–50.
Percival SL, Bowler PG, Russell D: Bacterial resistance to silver in wound care. J Hosp Infect 2005;60:1–7.

Poon VK, Burd A: In vitro cytotoxity of silver: implication for clinical wound care. Burns 2004;30: 140–147.

Qian L, Sun G: Durable and regenerable antimicrobial textiles: synthesis and applications of 3-methylol-2,2,5,5-tetramethylimidazolidin-4-one (MTMIO). J Appl Polym Sci 2003;89:2418–2425.

Qian L, Sun G: Durable and regenerable antimicrobial textiles: improving efficacy and durability of biocidal functions. J Appl Polym Sci 2004;91:2588–2593.

Randall LP, Cooles SW, Piddock LJ, Woodward MJ: Effect of triclosan or a phenolic farm disinfectant on the selection of antibiotic-resistant *Salmonella enterica*. J Antimicrob Chemother 2004;54: 621–627.

Räuchle A: Triclosan; in Kramer A, Weuffen W, Krasilnikow AP, Gröschel D, Bulka E, Rehn D (eds): Handbuch der Antiseptik. Antibakterielle, antifungielle und antivirale Antiseptik – Ausgewählte Wirkstoffe. Stuttgart, Fischer, 1987, vol II/3, pp 527–546.

Rayburn JR, Aladdin RK: Developmental toxicity of copper, chromium, and aluminum using the shrimp embryo teratogenesis assay: palaemonid with artificial seawater. Bull Environ Contam Toxicol 2003;71:481–488.

Reitsma AM, Rodeheaver GT: Effectiveness of a new antimicrobial gauze dressing as a bacterial barrier. 2001. http://www.kendallhq.com/catalog/ClinicalInformation/AntimicrobialBarrier.pdf

Riach CG, McBride D, O'Mailly ML: Triclosan: assessment of genotoxicity in an unscheduled DNA synthesis assay using adult rat hepatocyte primary cultures. Ciba Drug Master File, 1988.

Ricci G, Patrizi A, Bendandi B, Menna G, Varotti E, Masi M: Clinical effectiveness of a silk fabric in the treatment of atopic dermatitis. Br J Dermatol 2004;150:127–131.

Risbud MV, Karamuk E, Schlosser V, Mayer J: Hydrogel-coated textile scaffolds as candidate in liver tissue engineering. II. Evaluation of spheroid formation and viability of hepatocytes. J Biomater Sci Polym 2003;14:719–731.

Roed-Petersen J, Auken G, Hjorth N: Contact sensitivity to Irgasan DP 300. Contact Dermatitis 1975;1: 293–294.

Romagnani S: Human TH1 and TH2 subsets: regulation of differentiation and role in protection and immunopathology. Int Arch Allergy Immunol 1992;98:279–285.

Rothenburger S, Spangler D, Bhende S, Burkley D: In vitro antimicrobial evaluation of coated Vicryl Plus antibacterial suture (coated polyglactin 910 with triclosan) using zone of inhibition assays. Surg Infect 2002;3(suppl 1):79–88.

Rudolf M, Kampf G: Wirkstoffe; in Kampf G (ed): Hände-Hygiene im Gesundheitswesen. Berlin, Springer, 2003, pp 71–104.

Russell AD: Do biocides select for antibiotic resistance? J Pharm Pharmacol 2000;52:227–233.

Russell AD: Introduction of biocides into clinical practice and the impact on antibiotic-resistant bacteria. Symp Ser Soc Appl Microbiol 2002;31:121–135.

Russell AD: Whither triclosan? J Antimicrob Chemother 2003;53:693–695.

Russell AD, Maillard JY, Fuur JR: Possible link between bacterial resistance and use of antibiotics and biocides. Antimicrob Agents Chemother 1998;42:2151.

Russell LB, Montgomery CS: Use of the mouse spot test to investigate the mutagenic potential of triclosan (Irgasan DP300). Mutat Res 1980;29:7–12.

Salminen K, Talja R, Vasenius H, Weckström P: The fungous flora of the sauna and the influence of certain disinfectants. Zentralbl Bakt Hyg I Abt Orig B 1974;158:552–560.

Sanchez P, Moreno E, Martinez JL: The biocide triclosan selects *Stenotrophomonas maltophilia* mutants that overproduce the SmeDEF multidrug efflux pump. Antimicrob Agents Chemother 2005;49: 781–782.

Schäfer T: Ergebnisse einer randomisierten kontrollierten klinischen Studie zur Überprüfung der Wirksamkeit und des Tragekomforts einer silberbeschichteten Spezialtextilie bei atopischem Ekzem. In preparation.

Schäfer T, Staudt A, Ring J: Deutsches Instrument zur Erfassung der Lebensqualität bei Hauterkrankungen (DIELH). Hautarzt 2001;52:624–628.

Schauder S: Dermatologische Verträglichkeit von UV-Filtern, Duftstoffen und Konservierungsmitteln in Sonnenschutzpräparaten. Bundesgesundheitsbl Gesundheitsforsch Gesundheitsch 2001;44:471–479.

Schmid H, Dotti B, Keller B: 13-week oral toxicity (feeding) study with FAT 80'023/R (triclosan) in the hamster. Ciba Drug Master File, 1994.

Schmidt S, Fortnagel P, Wittich RM: Biodegradation and transformation of 4,4′- and 2,4-dihalodiphenyl ethers by *Sphingomonas* sp. strain SS33. Appl Environ Microbiol 1993;59:3931–3933.

Schmidt S, Wittich RM, Erdmann D, Wilkes H, Francke W, Fortnagel P: Biodegradation of diphenylether and its monohalogenated derivatives by *Sphingomonas* sp. strain SS3. Appl Environ Microbiol 1992;55:2744–2750.

Schmitt W: Wundinfektionen; in Schmitt W, Hartig W (eds): Allgemeine Chirurgie, ed 11. Leipzig, Barth, 1991, pp 556–572.

Schroder RE, Daly IW: A segment II teratology study in rats with Irgacare MP (C-P sample No 38328). Ciba Drug Master File, 1992.

Schultz A: Charakterisierung von extrazellulären Peroxidasen von Weissfäulepilzen der Gattungen *Fomitiporia* und *Nematoloma* und deren Bedeutung beim Abbau von Umweltschadstoffen; dissertation, Ernst-Moritz Arndt University, Greifswald, Math-Nat Fac, 2004.

Schweizer HP: Triclosan: a widely used biocide and its link to antibiotics. FEMS Microbiol Lett 2001;202:1–7.

Seebacher C, Kramer A: Pilzinfektionen in öffentlichen Sporteinrichtungen. Hautspektrum 1997;2:5–6.

Singer H, Muller S, Tixier C, Pillonel L: Triclosan: occurrence and fate of a widely used biocide in the aquatic environment – Field measurements in wastewater treatment plants, surface waters, and lake sediments. Environ Sci Technol 2002;36:4998–5004.

Sreenivasan P, Gaffar A: Antiplaque biocides and bacterial resistance: a review. J Clin Periodontol 2002;29:965–974.

Steinkjer B, Braathen LR: Contact dermatitis from triclosan (Irgasan DP 300). Contact Dermatitis 1988;18:243–244.

Storch M, Perry LC, Davidson JM, Ward JJ: A 28-day study of the coated Vicryl Plus antibacterial suture (coated polyglactin 910 suture with triclosan) on wound healing in guinea pig linear incisional skin wounds. Surg Infect 2002a;3(suppl 1):89–98.

Storch ML, Rothenburger SJ, Jacinto G: Experimental efficacy study of coated Vicryl Plus antibacterial suture in guinea pigs challenged with *Staphylococcus aureus*. Surg Infect 2004;5:281–288.

Storch M, Scalzo H, van Lue S, Jacinto G: Physical and functional comparison of coated Vicryl Plus antibacterial suture (coated polyglactin 910 suture with triclosan) with coated Vicryl suture (coated polyglactin 910 suture). Surg Infect 2002b;3(suppl 1):65–78.

Strachan DP: Hay fever, hygiene, and household size. BMJ 1989;299:1259–1260.

Suller MT, Russell AD: Antibiotic and biocide resistance in methicillin-resistant *Staphylococcus aureus* and vancomycin-resistant *Enterococcus*. J Hosp Infect 1999;43:281–291.

Suller MT, Russell AD: Triclosan and antibiotic resistance in *Staphylococcus aureus*. J Antimicrob Chemother 2000;46:11–18.

Sun G, Xu XJ, Bickett JR, Williams JF: Durable and regenerable antibacterial finishing of fabrics with a new hydantoin derivative. Ind Eng Chem Res 2001;40:1016–1021.

Sunderman FW, Lau TJ, Cralley LJ: Inhibitory effect of manganese on carcinogenicity of nickel subsulfide in rats. Cander Res 1974;34:92–95.

Takai K, Ohtsuka T, Senda Y, Nakao M, Yamamoto K, Matsuoka J, Hirai Y: Antibacterial properties of antimicrobial-finished textile products. Microbiol Immunol 2002;46:75–81.

Taplin D: Superficial mycoses. J Invest Dermatol 1976;67:177–181.

Thiemann L: Müssen moderne Textilien bei schweisstreibenden Tätigkeiten riechen und wie lässt sich dieses Problem beseitigen? 2004. http://www.baumann-online.de/ho_alois-kiessling/facharbeiten/Schweissgeruch/Schweissgeruch.htm.

Thom DC, Davies JE, Santerre JP, Friedman S: The hemolytic and cytotoxic properties of a zeolite-containing root filling material in vitro. Oral Surg Oral Med Oral Pathol Oral Radiol Endod 2003;95:101–108.

Thomann P, Maurer T: Skin sensitizing (contact allergenic) effect in guinea pigs of FAT 80023/A. Ciba Drug Master File, 1975.

Thurman RB, Gerba CHP: The molecular mechanisms of copper and silver ion disinfection of bacteria and viruses. Crit Rev Environ Contr 1989;18:295–315.

Tokura S, Nishimura SI, Sakairi N, Nishi N: Biological activities of biodegradable polysaccharide. Macromol Symp 1996;101:389–396.

Tulp MTM, Sundström G, Martron LBJM, Hutzinger O: Metabolism of chlorodiphenyl ethers and Irgasa® DP 300. Xenobiotica 1979;9:65–77.
Ugur A, Ceylan O: Occurrence of resistance to antibiotics, metals, and plasmids in clinical strains of *Staphylococcus* spp. Arch Med Res 2003;34:130–136.
Untiedt S: Deodorantien; in Umbach W (ed): Kosmetik und Hygiene. Weinheim, Wiley VCH, 2004, pp 358–368.
Vao R, Salkinoja-Salonen M: Microbial transformation of polychlorinated phenoxy phenols. J Gen Appl Microbiol 1986;32:505–517.
Veronesi S, de Padova MP, Vanni D, Melino M: Contact dermatitis to triclosan. Contact Dermatitis 1986;15:257–258.
Vigo TL: Protection of textiles from biological attack; in Lewin M, Sello SB (eds): Handbook of Fiber Science and Technology – Chemical Processing of Fibers and Fabrics, Functional Finishes Part A. New York, Dekker, 1983, vol II, p 367.
Vincent JL: Nosocomial infections in adult intensive-care units. Lancet 2003;361:2068–2077.
Voets JP, Pipyn P, van Lancker P, Verstraete W: Degradation of microbicides under different environmental conditions. J Appl Bacteriol 1976;40:67–72.
von Woedtke T, Schluter B, Pflegel P, Lindequist U, Jülich WD: Aspects of the antimicrobial efficacy of grapefruit seed extract and its relation to preservative substances contained. Pharmazie 1999;54:452–456.
Wahlberg JE: Routine patch testing with Irgasan DP 300. Contact Dermatitis 1976;2:292.
Wallhäusser KH: Praxis der Sterilisation: Desinfektion – Konservierung, ed 5. Stuttgart, Thieme, 1995.
Wallhäusser KH, Fischer K: Die antimikrobielle Ausrüstung von Textilien. Textilveredlung 1970;5:3–14.
Weber DJ, Rutala WA: Use of metals as microbiocides in preventing infections in healthcare; in Block SS (ed): Disinfection, Sterilization, and Preservation, ed 5. Philadelphia, Lippincott Williams & Wilkins, 2001, pp 415–430.
Werfel T: Staphylococcal toxin aggravates dermatitis in neurodermatitis. Krankenpfl J 2001;39:5–7.
Wigert H, Merka V, Cerovska M: Antimikrobielle Ausrüstung von Textilien; in Weuffen W (ed): Handbuch der Desinfektion und Sterilisation. Berlin, Volk und Gesundheit, 1974, vol III, pp 227–267.
Wilson BA, Smith VH, Denoyelles F, Larive CK: Effects of three pharmaceutical and personal care products on natural freshwater algal assemblages. Environ Sci Technol 2003;37:1713–1719.
Wnorowski G: Dermal sensitization test – Buehler method for triclosan lot No 5.2.0211.0. Ciba Drug Master File, 1994.
Wong CS, Beck MH: Allergic contact dermatitis from triclosan in antibacterial handwashes. Contact Dermatitis 2001;45:307.
Wuhrman KG, Zobrist F: Untersuchungen über die bakterizide Wirkung von Silber in Wasser. Mitt Eidgen Anst Wasserversorg Schweiz Hydrol 1958;20:218–255.
Xue C, Yu G, Hirata T, Terao J, Lin H: Antioxidative activities of several marine polysaccharides evaluated in a phosphatidylcholine-liposomal suspension and organic solvents. Biosci Biotechnol Biochem 1998;62:206–209.
Yau ET, Green JD: 2-year oral administration to rats – FAT 80'023: final report. Ciba Drug Master File, 1996.
Ye C, Zou H, Peng Y, Liu X, Chen Z: Preparation of chitosan-collagen sponge and its application in wound dressing. Sheng Wu Yi Xue Gong Cheng Xue Za Zhi 2004;21:259–260.
Zöllner H, Kramer A, Youssef P, Youssef U, Adrian V: Preliminary investigations on the biodegradability of selected microbicidal agents. Hyg Med 1995;20:401–407.

Prof. Dr. A. Kramer
Institute of Hygiene and Environmental Medicine, Ernst Moritz Arndt University
Walther-Rathenau-Strasse 49a
DE–17489 Greifswald (Germany)
Tel. +49 3834 515542, Fax +49 3834 515541, E-Mail kramer@uni-greifswald.de

Burg G (ed): Biofunctional Textiles and the Skin.
Curr Probl Dermatol. Basel, Karger, 2006, vol 33, pp 110–126

Production Process of a New Cellulosic Fiber with Antimicrobial Properties

Stefan Zikeli

Zimmer AG, Frankfurt/Main, Germany

Abstract

The Lyocell process (system: cellulose-water-N-methylmorpholine oxide) of Zimmer AG offers special advantages for the production of cellulose fibers. The process excels by dissolving the most diverse cellulose types as these are optimally adjusted to the process by applying different pretreatment methods. Based on this stable process, Zimmer AG's objective is to impart to the Lyocell fiber additional value to improve quality of life and thus to tap new markets for the product. Thanks to the specific incorporation of seaweed, the process allows to produce cellulose Lyocell fibers with additional and new features. They are activated in a further step – by specific charging with metal ions – in order to obtain antibacterial properties. The favorable textile properties of fibers produced by the Lyocell process are not adversely affected by the incorporation of seaweed material or by activation to obtain an antibacterial fiber so that current textile products can be made from the fibers thus produced. The antibacterial effect is achieved by metal ion activation of the Lyocell fibers with incorporated seaweed, which contrasts with the antibacterial fibers known so far. Antibacterial fibers produced by conventional methods are in part only surface finished with antibacterially active chemicals or else they are produced by incorporating organic substances with antibacterial and fungicidal effects. Being made from cellulose, the antibacterial Lyocell fiber Sea Cell® Active as the basis for quality textiles exhibits a special wear comfort compared to synthetic fibers with antibacterial properties and effects. This justifies the conclusion that the Zimmer Lyocell process provides genuine value added and that it is a springboard for further applications.

Zimmer AG is an innovative and technology-oriented engineering contractor in the field of 'man-made' fiber production.

After several years of research and development activities, Zimmer AG is launching a new, multifunctional cellulosic fiber with bioactive properties.

In response to the continuously rising demand for antibacterial fibers, this innovative, high-tech fiber is perfectly suited for all applications where hygiene and cleanliness are required.

The antibacterial effect constitutes an integrated, permanent component of the cellulosic fiber matrix which remains unaffected by wear, e.g. washing or dry-cleaning.

The fiber protected worldwide by industrial property rights has proven its activity against the most diverse types of bacteria.

Bacteria develop and spread wherever they find a growth-stimulating combination of heat and humidity.

In particular for sportswear and leisure wear but also for bed linens, underwear or filling materials, measures that prevent or reduce bacterial growth constitute an advantage.

The antibacterial effect of many conventional fibers is achieved by way of organic compounds known from the pesticide industry. Such compounds are either incorporated into the fiber or applied to its surface.

These substances, which often have secondary effects that should not be neglected, may cause skin irritations. One of the objectives of the research and development activities of Zimmer AG was to achieve a bactericidal effect without any negative impact on the human body caused by organic compounds, if possible.

The permanent bactericidal effect combined with the beneficial effect on the skin that is typical of cellulosic fibers was examined and confirmed by numerous tests conducted at the Hohenstein Research Institute, among others.

The results obtained so far regarding the textile processability have shown that the new fiber can be combined with the most varied textile fibers and natural substances and that it is thus universally applicable to all cases where a bactericidal combined with a beneficial effect on the skin is required.

This new cellulosic, antimicrobial fiber is produced by Sea Cell GmbH in Rudolstadt, Germany, a 100% subsidiary of Zimmer AG.

Conventional Bioactive Fibers and Their Active Substances

The Substances

Antibacterial substances are chemicals that inhibit bacterial growth and/or destroy bacteria.

Depending on their mode of action, a difference is made between bactericidal substances which destroy bacteria and bacteriostatic substances which only inhibit the growth of bacteria.

Table 1. Commonly known bioactive substances [1–3]

Organic active ingredients	Halogenated diphenyl ethers (e.g. triclosan) Phenolic compounds Halophenols and bisphenolic compounds Resorcinol and its derivatives Benzoic esters Quaternary ammonium compounds
Metals	Silver, zinc, copper
Other anorganic ingredients	Zeolites NaAl silicate

To this effect, a multitude of bioactive organic, metallic and other inorganic substances are commonly known and available (table 1) [1–3].

To obtain the desired antibacterial effect, the individual substances or a combination of several ingredients are used depending on the respective requirements and application. One example is a combination of zeolites and metals such as silver, zinc or copper [4, 5].

Antibacterial Finishing of Synthetic Fibers

The following options are available to manufacture antibacterial fibers:

- incorporation of antibacterial substances into the spinning solution;
- application and fixing of the antibacterial substance on the surface of the fiber in an aftertreatment step;
- modification of the polymer and application of the bioactive substance.

Conventional antibacterial fibers on the basis of synthetic polymers are listed in table 2.

One widely used substance in antibacterial fibers is triclosan, a halogenated bisphenyl ether that is mainly used as a pesticide but can also be found as an additive in soaps or toothpaste.

The use of triclosan is controversial, not only because of its application as a pesticide, but also with a view to its production. In the synthesis of triclosan, carcinogenic substances such as dioxins and dibenzofurans can appear as byproducts.

Triclosan is contained as an active ingredient for example in Microban® products as well as in the polyacryl fiber Amicor® by Acordis.

Antibacterial Finishing of Cellulosic Fibers

Viscose fibers with an antibacterial effect are common on the fiber market [3].

The bactericidal effect is obtained by adding inorganic compounds during the spinning process.

Table 2. Conventional antibacterial fibers on the basis of synthetic polymers

Polymer	Company	Brand
Polyester	Trevira	Trevira Bioactive
	Montefibre	Terital Saniwear
	Brilen-Nurel	Bacterbril
	Dupont-Sabanci	Allerban
	Kanebo	Bactekiller
	Kuraray	SA30 and SA21
	Wellman	Filwell Wellcare
Polyacryl	Acordis	Amicor
	Sterling	Biofresh
	Montefibre	Leacril Saniwear
Polyamide	Kanebo	Livefresh
	R-STAT	R-STAT
	Unitika	Bioliner
	Nylstar	Meryl Skinlife
Polypropylene	Drake	Permafresh
	Asota	Asota AM Sanitary
Polyvinylchloride	Rhovyl	Rhovyl's as antibacterial

For the manufacture of fibers such as antibacterial cellulose acetate fibers, just like synthetic fibers, mainly the substance triclosan is used. In this case, the substance is incorporated into the core of the fiber and is released through its surface. A well-known fiber of this type is Microsafe AM by Hoechst/Celanese, for example.

A particularity in the field of cellulosic fibers is the antibacterial olynosic fiber offered by Fuji – Chitopoly – which is manufactured by adding chitin during fiber production.

Lyocell fibers with antibacterial finishing are unknown so far.

Manufacture of Sea Cell® Active Fibers

The Sea Cell Process

The Lyocell process has established itself as an environment-friendly, economically viable, product-enhancing and highly flexible alternative for the manufacture of man-made cellulose fibers. It is continuously improved and marketed by Zimmer AG and Sea Cell GmbH.

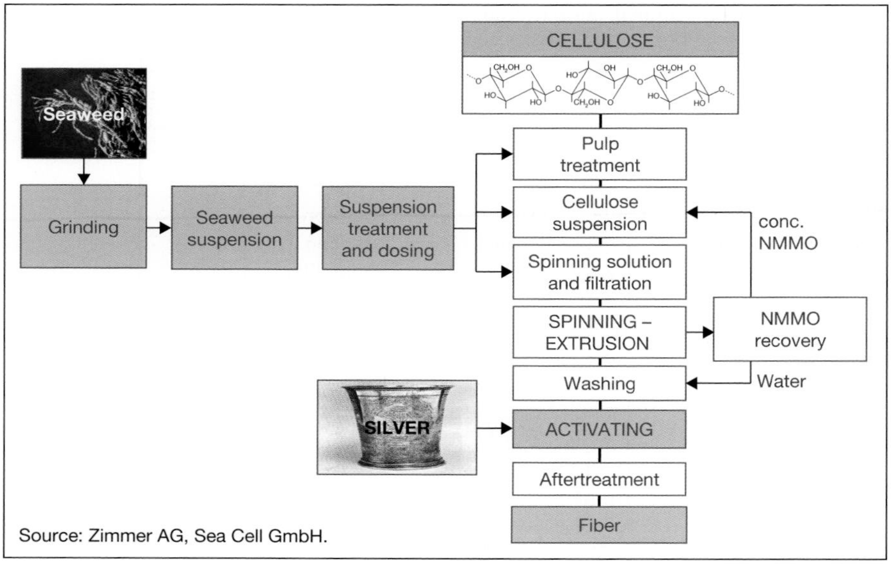

Fig. 1. Sea Cell Active – production process. NMMO = N-methylmorpholine-N-oxide.

In the Lyocell process, the cellulose is dissolved directly without formation of derivatives [6, 7]. As a solvent, the process uses the nontoxic, aqueous solvent N-methylmorpholine-N-oxide.

The spinning solution is processed in a combined dry/wet spinning process (air gap) to form fibers and shaped cellulose articles [8, 9]. During the spinning process, the solvent required to produce the spinning solution is washed out and almost completely recovered.

The Sea Cell fibers are manufactured (fig. 1) by adding finely ground seaweed, mainly from the family of brown, red, green and blue algae, but particularly the brown alga *Ascophyllum nodosum* and/or the red alga *Lithothamnium calcareum*, to the spinning solution [10–14].

The algae are added either as a powder or as a suspension in one of the process steps preceding the spinning of the cellulose solution.

Seaweed has the capability of absorbing the minerals contained in seawater. An analysis showed that in addition to minerals also carbohydrates, amino acids, fats and vitamins are found in the seaweed.

Given this variety of active ingredients, seaweed and/or seaweed extracts are preferably used in cosmetics as well as in the pharmaceutical industry [15–19].

In addition to the special properties afforded by seaweed, the Sea Cell fiber also exhibits a remarkably high tensile strength in dry and wet conditions as well as negligible shrinking. Based on the good physical properties of the textiles, fabrics made from Sea Cell fibers offer high dimensional stability in addition to high wear comfort which is typical of cellulosic Sea Cell fibers and fiber mixtures.

Metal Absorption of the Sea Cell Fiber

One particularity of the Sea Cell fiber is its capacity to bind and absorb substances. During the activation of pure Sea Cell fibers, bactericidal metals like silver, zinc, copper and many others are absorbed by the fully formed cellulosic fiber through metal sorption.

Unlike the commonly used method of incorporating the active ingredients in the spinning solution, the manufacture of Sea Cell Active offers the possibility of incorporating the substance permanently into the core of the fully formed fiber in an activation step.

Impregnation tests with diluted metal salt solutions surprisingly showed that the Sea Cell fiber exhibits excellent sorptive behavior regarding metals and/or metal ions.

It must be assumed that the metals are bonded via free carbonyl, carboxy and hydroxyl groups of the cellulose as well as of the incorporated seaweed. It is well known that phenols contained in the seaweed have the ability to chelate heavy metals [20, 21].

The metal ions are firmly anchored in the fiber matrix through the swelling of the cellulose which promotes an even distribution of the seaweed over the fiber cross-section.

Even conventional cleaning methods that usually require an alkaline atmosphere do not affect the metal concentrations in the loaded Sea Cell fiber.

To obtain fibers with a permanent antibacterial finishing, ionic silver and zinc solutions are used for loading/activating.

This constitutes an important advantage compared to the antibacterial fibers with surface treatment available on the market.

Antibacterial Properties of Zinc and Silver

Zinc offers an antibacterial and anti-inflammatory effect and can be used to combat bacteria (like streptococci and *Actinomyces*) [22].

The antibacterial effect of silver was already known in ancient times. Silver tools and containers (approx. 4,000 BC) were used for storing and transporting water to prevent the formation of germs and ensure high water quality [23].

In the 19th century it was evidenced that silver has an antimicrobial effect even in smallest concentrations and that it quickly destroys typhoid bacilli and the resistant anthrax spores at concentrations as low as 1:4,000 to 1:10,000 [23].

In the decontamination of water as well as in the disinfection of wounds, the oligodynamic effect of silver (inhibition of bacterial growth already at metal

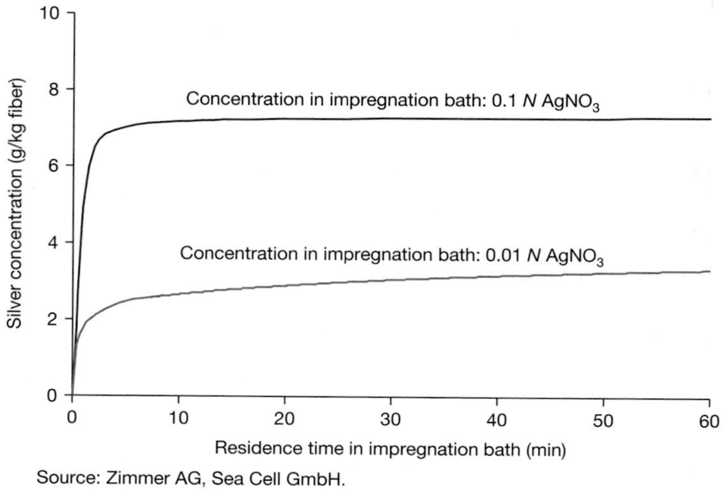

Fig. 2. Sea Cell pure fiber – absorption of silver.

concentrations of 0.006–0.5 ppm) is used [23, 24]. Bedding and textiles with minor concentrations of silver already show positive effects in the treatment of neurodermatitis or psoriasis [25].

As silver does not have any negative side effects like skin irritation, this metal is favored for use as an activating agent to manufacture Sea Cell Active fibers [24, 26].

As a consequence of the very low but sufficient leaching out of silver ions from the Sea Cell Active fiber, the antibacterial activity remains unchanged over the long term. From a dermatological and hygienic viewpoint, it is therefore ensured that no skin irritation may occur. The excellent wear comfort of the cellulosic fiber is not affected.

Absorption Kinetics

The Sea Cell Active fiber is produced by loading the pure Sea Cell fiber with a diluted silver solution. The absorption of silver constitutes the decisive process step for the manufacture of this fiber which is why the absorption kinetics were examined as a function of the silver concentration in the bath solution.

Figure 2 shows the absorption kinetics of 0.1 and 0.01 *N* silver nitrate solution. The figures show that the loading process is largely completed after a retention time of 1–3 min. A more important factor than the retention time for the absorption of silver by the fiber is the bath concentration. When treated

Table 3. Main minerals contained (mg/kg fiber) in Lyocell, pure Sea Cell and Sea Cell Active

Minerals	Lyocell	Pure Sea Cell	Sea Cell Active
Silver	–	–	6,900
Calcium	38	1,800	1,540
Magnesium	95	275	107
Sodium	306	330	13

with 0.1 N $AgNO_3$ bath solution, approximately 7 g of silver were absorbed per kilogram of Sea Cell fiber while in a 0.01 N $AgNO_3$ bath solution, approximately 3.5 g/kg of fiber were absorbed.

Chemical Analysis of the Sea Cell Active Fiber

To find out what effect the activation step has on the composition of the fiber, the main minerals contained in the Sea Cell Active fiber were analyzed and compared with those of the Lyocell and regular Sea Cell fiber.

Table 3 shows that the absorbed silver reduces the concentrations of alkali and earth alkali metals. This is most evident in the decline of the sodium concentration.

In addition to atomic absorption spectroscopy, a quantitative and wet chemical analysis, the distribution of silver over the cross-section of the fiber was examined. To this effect, the Sea Cell Active fiber was broken using liquid nitrogen. This cryogenic break was examined with a scanning electron microscope (Leo type DSM 962). In addition to the backscattering picture, the distribution of silver at the break surface was examined using energy-dispersive X-ray analysis (EDX detector by Oxford Instruments).

The picture in figure 3 compares the electron backscattering picture with the corresponding energy-dispersive X-ray mapping.

In the backscattering picture, which also shows the material contrast (fig. 3a), the darker spots indicate light elements while the light spots indicate heavy elements like silver.

In the silver mapping pictures (fig. 3b), the light spots indicate the presence of silver.

These pictures clearly show that the silver is distributed evenly across the fiber cross-section and that it is not only exclusively found at the surface of the fiber.

Leaching Tests

To test how durably silver has been incorporated into the Sea Cell Active fiber, a raw fabric made from 70% cotton, 20% pure Sea Cell and 10% Sea Cell Active was extracted in a Soxleth apparatus.

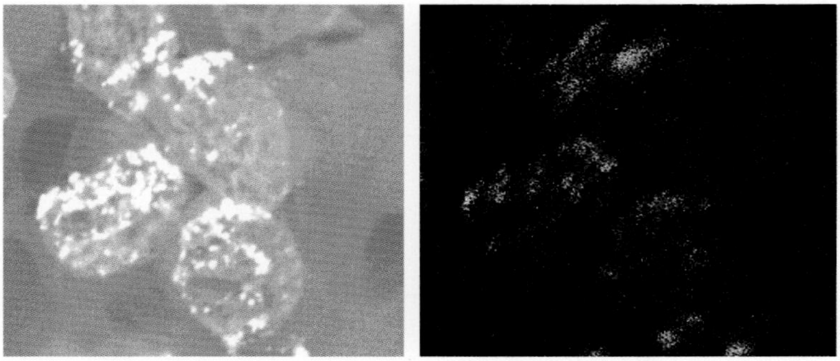

a b

Method: Cryogenic break of the fibers Scanning electron microscope (Leo, type DSM 962) Energy-dispersive X-ray spectroscopy Oxford instruments

Source: ITCF Denkendorf.

Fig. 3. Sea Cell Active – silver distribution. ***a*** Scanning electron-microscopic image. ***b*** Energy-dispersive X-ray silver mapping.

The extraction of raw fabric was achieved using distilled water. A sample from the total extraction liquid was taken after each cycle.

After the extractions the extracted raw fabric was dried in a circulating hot air drying cabinet.

The analyses showed that the raw fabric contained the same amount of silver prior to and after the extraction. The aqueous extract only showed silver concentrations close to the detection limit. This test proves that no silver leaches from the raw fabric in an extraction with distilled water.

Fields of Application of Sea Cell Active

The Sea Cell Active fiber as an antibacterial fiber for fabrics and nonwoven materials can be used as a pure, i.e. 100% fiber or in fiber blends.

Thus, the product is suited for a multitude of applications:

- working wear (incl. gloves)
- sportswear (incl. socks)
- underwear and lingerie
- home textiles (furnishing and bedding, fillers)
- nonwoven materials and technical applications (wipes, filters and masks)
- household and hygienic applications.

Table 4. Tenacity and elongation properties

	Lyocell	Pure Sea Cell	Sea Cell Active
Titer (typical), dtex	1.3	1.4	1.4
Tenacity cond., cN/tex	36.5	35.9	34.4
Tenacity wet, cN/tex	31.4	31.1	2.8
Elongation cond., %	12.1	11.9	9.3
Elongation wet, %	15.3	13.4	14.2

Physical Properties of the Sea Cell Active Fiber

Sea Cell Active fibers can be produced as staple fibers and as filament products in all fiber diameters and cut lengths. Experience has shown that the high tenacity and elongation properties remain unchanged in the production of the Sea Cell Active fibers (table 4).

Yarns can be easily produced with either 100% Sea Cell Active or as blends in all standard counts.

Antimicrobial Properties of the Sea Cell Active Fiber

With the above we have proven that silver as an antimicrobial substance can be incorporated into the Sea Cell fiber. The question is now whether the absorbed silver shows the expected antimicrobial activity.

Antibacterial Activity – Testing Method

The antibacterial activity was determined following the Japanese Industrial Standard (JIS L1902-1998 'Testing method for antibacterial activity of textiles'). The test was performed using the Gram-positive strain of *Staphylococcus aureus* (ATCC 6538P) and the Gram-negative strain of *Klebsiella pneumoniae* (DSM 789).

The germs were applied to the fibers to be analyzed and to reference samples. The germ count (the number of colony-forming units, CFU) on the fibers and on the reference sample was determined after an 18-hour incubation period. The difference between the germ count of the fiber and that of the reference sample shall be the parameter for the germ-inhibiting and/or germicidal activity of the fiber under analysis.

The antibacterial activity is divided into three groups: slight activity, significant activity and strong activity.

A slight antibacterial activity is found where the difference of the CFU log between the fiber and the reference sample equals zero, a significant activity is

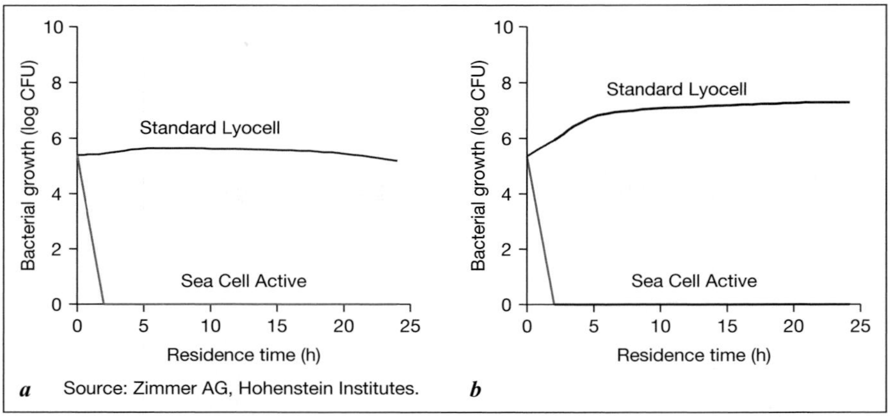

Fig. 4. Bacterial growth on sea Cell Active versus Lyocell. ***a*** *Staphylococcus aureus*. ***b*** *Klebsiella pneumoniae*.

Table 5. Specific efficacy (log CFU)

Antibacterial activity	Specific efficacy
Slight	0 ± 0.5
Significant	>1 ± 0.5
Strong	>3 ± 0.5

found where the difference is greater than 1 and a strong activity is found where the difference is greater than 3.

This parameter is also referred to as the specific efficacy (table 5).

Antibacterial Activity – Sea Cell Active versus Lyocell

In figure 4, the growth of *Staphylococcus* and *Klebsiella* on a standard Lyocell fiber and on a Sea Cell Active fiber is shown as a function of the residence time.

Unlike the standard Lyocell fiber where the bacterial growth is measurable, the Sea Cell Active fiber exhibits a strong bactericidal effect. Within as little as 2 h, no more bacteria could be detected on the fiber.

Antibacterial Activity of Sea Cell Active in Fiber Blends

The examination of the antibacterial activity was not only conducted with pure Sea Cell Active products but also using blends with Lyocell. In all blends (table 6), a strong antibacterial activity was measured.

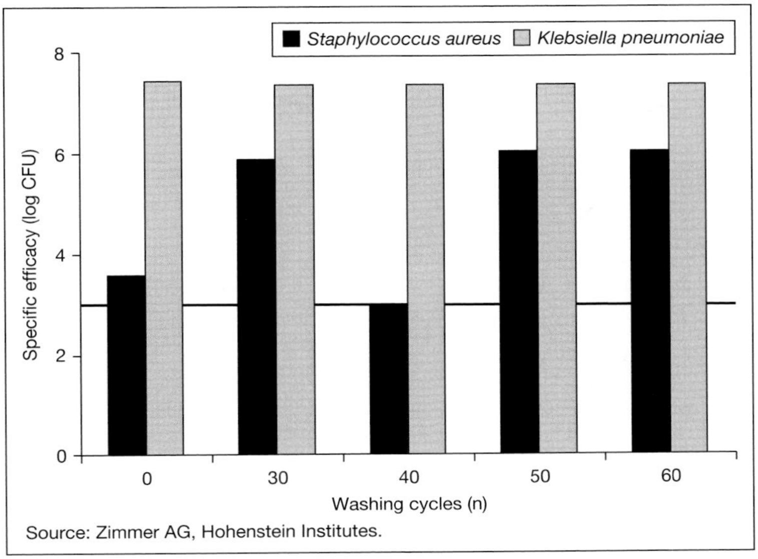

Fig. 5. Sea Cell Active – antibacterial durability.

Table 6. Antibacterial activity in fiber blends

	Specific efficacy, log CFU		Antibacterial activity
	S. aureus	*K. pneumoniae*	
Sea Cell Active	5.79	7.54	strong
Blend 1 (Sea Cell Active/ Lyocell, 25/75%)	5.79	7.54	strong
Blend 2 (Sea Cell Active/ Lyocell, 15/85%)	5.79	7.54	strong

Permanence of the Antibacterial Activity of Sea Cell Active

To determine the permanence of the antibacterial activity of Sea Cell Active fibers, the samples were washed 60 times following the specifications of DIN ISO 6330 (5a – standard washing cycle at 40°C).

Figure 5 shows the antibacterial activity (specific efficacy) of the Sea Cell Active fibers. A strong antibacterial impact is detected at log CFU >3.

Despite 60 washing cycles, the reduction in the silver concentration of the fiber remains negligible.

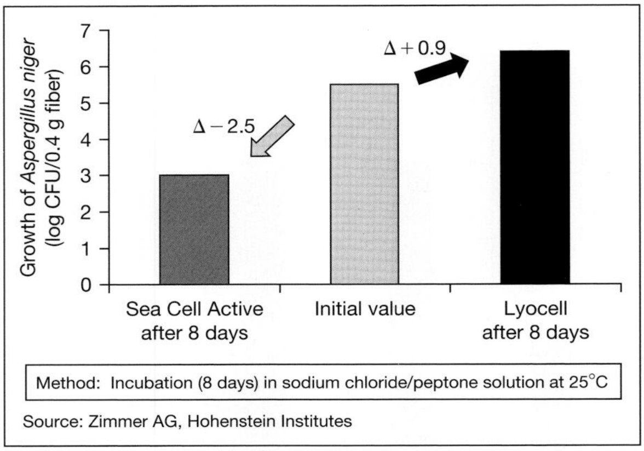

Fig. 6. Sea Cell Active – fungicidal impact.

The strong antibacterial activity against *Staphylococcus* and *Klebsiella* remains unchanged even after 60 washing cycles.

Fungicidal Effect

In addition to the antibacterial activity, also the fungicidal effect of Sea Cell Active was examined.

Tests to evidence the fungicidal activity were conducted using an *Aspergillus niger* DSM 1957 test culture.

To prove the fungicidal effect of our innovative fiber, the spore suspension of *A. niger* was applied to both Sea Cell Active and the reference Lyocell sample and incubated for 8 days. The colony count was determined at the start and at the end of the test.

While the concentration of *A. niger* after 8 days of incubation rose by 0.9 log steps on the reference Lyocell fiber, it dropped by 2.5 log steps on the Sea Cell Active fiber (fig. 6).

These tests clearly prove the fungicidal effect of Sea Cell Active fibers.

Bio- and Skin Compatibility

Zimmer AG's Lyocell process constitutes the environmentally friendly baseline technology and has been awarded the European Union's eco label. Only inoffensive natural components are used for the production of Sea Cell Active fibers, namely cellulose, seaweed and silver. Sea Cell Active has been certified according to the Ökotex Standard 100 for the product class 1 (baby articles). Furthermore extensive testing for bio- and skin compatibility has been carried out.

Cytotoxicity

The cytotoxicity test is a standardized and sensitive biocompatible testing method used for the assessment of materials intended for use in medical applications. The test serves to differentiate between reactive and nonreactive materials and provides information on the biocompatibility of materials.

Pursuant to ISO 10993–5, Cytox prepared extracts of pure Sea Cell and Sea Cell Active in different dilutions (90, 30, 10, 3.3%). These extracts as well as a positive control sample (toxic Trition) and a negative control sample (nontoxic nutrient culture) were applied to precultured L929 mouse fibroblast cells. After an incubation time of 48 h (37°C, 5% pCO_2) the protein concentrations in the cells were determined.

Materials whose extracts result in a reduction of the protein content of L929 cells by more than 19% in comparison with the negative control sample (nontoxic nutrient culture) are rated as cytotoxic.

This was not the case neither with the extracts of pure Sea Cell nor with the extracts of Sea Cell Active. Therefore the pure Sea Cell and Sea Cell Active fibers are not cytotoxic and comply with the requirements of the cytotoxicity test pursuant to ISO 10993–5 (1999).

Skin Compatibility Test with Human Keratinocytes

While the cytotoxicity test examines whether the fiber releases toxic and thus strongly irritating substances, the cytokine test using keratinocytes is more sensitive and also detects weakly irritating substances. The tests were conducted by ITVP–ITV Denkendorf Produktservice GmbH, Germany, applying a test procedure developed by the dermatological clinic of Heidelberg University, Germany. The advantage of this keratinocyte test is that it does not detect individual substances or substance classes but that it serves to assess the skin compatibility in general on the basis of the effect.

The two fibers, pure Sea Cell and Sea Cell Active, were analyzed as to their skin compatibility by determining the cytokines relevant for inflammation using human keratinocytes: to this effect, the fibers were incubated for 16 h at 37°C in a special transfer system to simulate a prolonged skin contact under wearing conditions. The extract of the transfer was γ-sterilized and incubated for 24 h with a confluent cell culture of human keratinocytes. The supernatant of the keratinocyte culture contained 4 different inflammation mediators (IL-1α, IL-6, IL-8 and GM-CSF) – these are messenger substances which may cause inflammation.

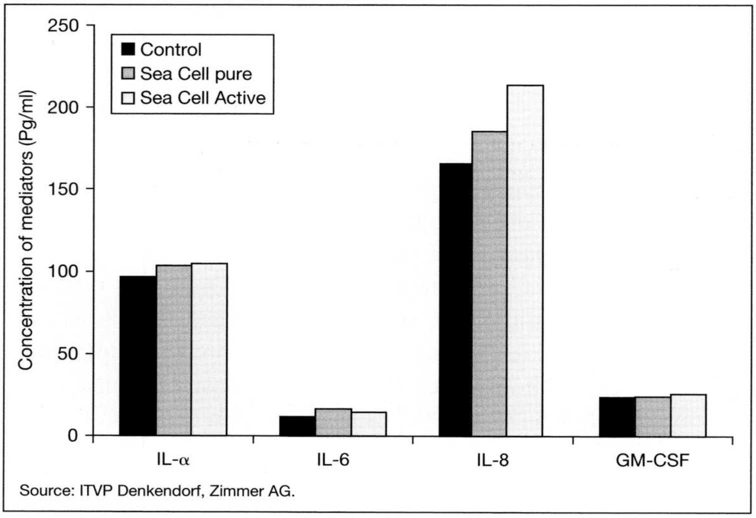

Fig. 7. Sea Cell Active – skin compatibility.

A detectable irritating effect is found if at least one of the mediators shows a 50% higher value than the control value.

Pure Sea Cell and Sea Cell Active showed no detectable irritating effect. The fibers proved to be skin compatible in the keratinocyte test (fig. 7).

Conclusion

Process

Zimmer AG's Lyocell process constitutes the environmentally friendly baseline technology and has been awarded the European Union's eco label.

Safe, natural components like cellulose, seaweed and silver are processed.

Activation with silver (metal sorption) is easy from a technical point of view and bears no health risk.

The production process does not require handling of pesticides or pesticide-like substances.

Product

Demand is rising for fibers with antibacterial effect.

The organic compounds used so far (pesticides) are controversial in terms of application (skin irritation) and production.

Sea Cell Active uses silver as a safe active substance without the need for organic compounds, and, therefore, the skin compatibility of the cellulose fiber is not impaired.

Sea Cell Active is the first cellulosic Lyocell fiber with an antimicrobial effect; it shows both a bactericidal and a fungicidal activity.

Sea Cell Active acts permanently, even after frequent washing.

The particular properties of cellulosic fibers (wear comfort) remain unchanged in the Sea Cell Active fiber.

The excellent physical properties are not affected by the addition of silver.

Application

There is universal applicability because textiles can be made from both 100% Sea Cell Active material or from blends with any type of fiber while still retaining its antibacterial activity.

Given its miscibility with other fibers, Sea Cell Active is suited for a broad range of applications in work wear and sportswear, underwear, home textiles and technical applications.

Acknowledgements

The author would like to extend his special thanks to the Institut Hohenstein, ICF Denkendorf, Sea Cell GmbH, Lurgi Austria GmbH as well as Zimmer AG.

References

1 Cianciolo AD, Lines EL: Polyurethane foam having cleaning and germicidal activities. Patent US 4,476,251. October 9, 1984.
2 Beerse PW, Morgan JM, Baier KG, Cen W, Bakken TA: Antimicrobial wipes which provide improved residual benefit versus gram negative bacteria. Patent US 6,197,315. March 6, 2001.
3 Vigo TL: Antibacterial fibers; in Rowell RM (ed): Modified Cellulosics. London, Academic Press, 1978, pp 259–284.
4 Uchida A, Ito H: Viscose sponge having antibacterial property. Patent JP2153723. June 13, 1990.
5 Mans L, Hammer KD: Sponge towel based on cellulose and a method for producing the same. Patent WO99/27835. June 10, 1999.
6 Zikeli S, Wolschner B, Eichinger D, Jurkovic R, Firgo H: Process for producing solutions of cellulose. Patent EP 0356419. December 16, 1992.
7 Bauer R, Taeger E, Zikeli S: Alceru – Manufacture of cellulosic materials by using organic solvent process. Seminar Cellulosic Man-Made Fibers in the New Millennium, Stenungsund, June 2000.
8 Zikeli S, Ecker F: Method for spinning a spinning solution and spinning head. Patent WO 01/81663. November 1, 2001.
9 Zikeli S, Ecker F: Verfahren und Vorrichtung zur Herstellung von Endlosformkörpern. Patent DE10037923. March 29, 2001.
10 Zikeli S: Lyocell fibers with health-promoting effect through incorporation of seaweed. Chem Fibers Int 2001;51:272–276.

11 Zikeli S: Lyocell-Fasern mit gesundheitsfördernder Wirkung durch Inkorporation von Algen. 7. Symp Nachwachsende Rohstoffe für die Chemie, Dresden, Germany, 2001, vol 18, pp 449–466.
12 Zikeli S, Endl T, Martl MG: Polymer compositions and moulded bodies made therefrom. Patent WO01/62844. August 30, 2001.
13 Zikeli S, Endl T, Martl MG: Cellulose sharped body and method for the production thereof. Patent WO01/62845. August 30, 2001.
14 Strasburger E, Noll F, Schenck H, Schimper AFW: Lehrbuch der Botanik für Hochschulen, ed 33. Stuttgart, Fischer, 1991.
15 Hoppe HA, Levering T, Tanaka Y: Marine Algae in Pharmaceutical Science. Berlin, de Gruyter & Co, 1979.
16 Vasage M, Rolfsen W, Bohlin L: Sulpholipid composition and methods for treating skin disorders. Patent US 6,124,266. September 26, 2000.
17 Hills CB: Extraction of anti-mutagenic pigments from algae and vegetables. Patent US 4,851,339. July 25, 1989.
18 Ruegg R: Extraction process for beta carotene. Patent US 4,439,629. March 27, 1984.
19 Henrikson R: Earth Food Spirulina, ed 5. Kenwood, Ronore Enterprises, 1999.
20 Pedersen A: Studies on phenol content and heavy metal uptake in fucoids. Hydrobiologia 1984;116/117:498–504.
21 Leusch A, Holan ZR, Volesky B: Biosorption of heavy metals by chemically reinforced biomass of marine algae. J Chem Tech Biotechnol 1995;62:279–288.
22 Eby GA: Handbook for Curing the Common Cold – The Zinc Lozenge Story. Austin, Tex., USA, George Eby Research, 1994.
23 Grier N: Silver and its compounds; in Block SS (ed): Antiseptics and Disinfectants. Philadelphia, Lea & Febiger, 1977, pp 375–389.
24 Lansdown ABG: Silver I: its antibacterial properties and mechanism of action. J Wound Care 200;11:125–130.
25 Gauger A, Mempel M, Schekatz A, Schäfer T, Ring J, Abeck D: Silver-coated textiles reduce *Staphylococcus aureus* colonization in patients with atopic eczema. Dermatology 2003;207:15–21.
26 Haynes JL, Schulte TH: Antibacterial silver surfaces – An assessment of needs and opportunities for clinical devices. First Int Conf on Gold and Silver in Medicine, Marylena, May 1987.

Dr. Stefan Zikeli
Zimmer AG
Borsigallee 1
DE–60388 Frankfurt am Main (Germany)
Tel. +43 69 4007 222, Fax +43 69 4007 766

Office Austria
Wartenburgerstrasse 1a
AT–4840 Vöcklabruck (Austria)
Tel. +43 7672 24522 70, Fax +43 7672 24522 10, E-Mail zikeli@seacell.com

Burg G (ed): Biofunctional Textiles and the Skin.
Curr Probl Dermatol. Basel, Karger, 2006, vol 33, pp 127–143

Use of Textiles in Atopic Dermatitis

Care of Atopic Dermatitis

G. Ricci[a], *A. Patrizi*[b], *F. Bellini*[a], *M. Medri*[b]

[a]Department of Paediatrics and [b]Department of Clinical and Experimental Medicine, Division of Dermatology, University of Bologna, Bologna, Italy

Abstract

Atopic dermatitis (AD) is a chronic relapsing inflammatory skin disease which usually starts during the first years of life. In the management of AD, the correct approach requires a combination of multiple treatments to identify and eliminate trigger factors, and to improve the alteration of the skin barrier. In this article we try to explain the importance of skin care in the management of AD in relation to the use of textiles: they may be useful to improve disrupted skin but they are also a possible cause of triggering or worsening the lesions. Garments are in direct contact with the skin all day long, and for this reason it is important to carefully choose suitable fabrics in atopic subjects who have disrupted skin. Owing to their hygienic properties fabrics produced from natural fibres are preferential. Wool fibres are frequently used in human clothes but are irritant in direct contact with the skin. Wool fibre has frequently been shown to be irritant to the skin of atopic patients, and for this reason wool intolerance was included as a *minor criterion* in the diagnostic criteria of AD by Hanifin and Rajka in 1980. Cotton is the most commonly used textile for patients with AD; it has wide acceptability as clothing material because of its natural abundance and inherent properties like good folding endurance, better conduction of heat, easy dyeability and excellent moisture absorption. Silk fabrics help to maintain the body temperature by reducing the excessive sweating and moisture loss that can worsen xerosis. However, the type of silk fabric generally used for clothes is not particularly useful in the care and dressing of children with AD since it reduces transpiration and may cause discomfort when in direct contact with the skin. A new type of silk fabric made of transpiring and slightly elastic woven silk is now commercially available (Microair Dermasilk®) and may be used for the skin care of children with AD. The presence of increased bacterial colonization has been demonstrated in patients with AD. Such colonization has been included in the group of trigger factors for eczema in AD. Silver products have recently been demonstrated to offer two advantages in the control of bacterial infections. Textiles may be used not only for clothes, but also to prevent dust mite sensitization in atopic patients. A marked clinical improvement of AD was observed in a group of adults and children with positive skin tests (not necessarily towards mites), after an intensive eradication programme for mite allergens. Skin treatment with acaricide and house dust mite control measures can decrease AD symptoms. Different textiles have various potential worsening links with allergies: e.g. clothing has been proposed as an additional source of exposure to mite and cat allergens. On the other hand, special textiles can be used to prevent dust mite sensitization.

Atopic dermatitis (AD) is a chronic relapsing inflammatory skin disease which usually starts during the first years of life. There are many factors known to worsen the disease, including food and inhalant allergens, climatic factors, skin infections due to *Staphylococcus aureus*, stress, chemical and physical irritants. In the management of AD, the correct approach requires a combination of multiple treatments to identify and eliminate trigger factors, and to improve the alteration of the skin barrier. Dressings can be an effective barrier against persistent scratching, allowing more rapid improvement of the lesions, and may reduce the external source of skin irritants and bacterial infection. However, the rough fibres of some fabrics used as dressings may produce a worsening in skin conditions. Acute and cumulative irritation, allergic contact dermatitis, exacerbation of AD and contact urticaria have been reported to have been caused by textile fibres (e.g. nylon can cause allergic contact dermatitis and contact urticaria, wool can cause acute and cumulative irritant dermatitis and aggravate AD). In the following article we try to explain the importance of skin care in the management of AD in relation to the use of textiles: they may be useful to improve disrupted skin but they are also a possible cause of triggering or worsening the lesions.

The Skin of Atopic Patients: Physiopathology

Skin provides the barrier between the body and the environment and the first line of defence against exogenous noxious agents. The stratum corneum, the human main permeability barrier, is formed from extracellular lipids and corneocytes which are the result of the epidermal differentiation of the skin. Epidermal lipids are synthesized by keratinocytes and stored in epidermal lamellar bodies known as membrane-coating granules, lamellar granules or Odland bodies, which contain free sterols nearly all present as cholesterol, phospholipids and glycolipids, lipases and glycosidases and hydrolytic enzymes including β-glucocerebrosidase and acid and neutral sphingomyelinases [1]. These organelles are adjacent to the cellular membrane and fuse with the plasma membrane dispersing their content into the interstices. The hydrolytic enzymes are delivered to the intercellular space of the stratum corneum where they convert secreted glucosylceramides and phospholipids, including sphingomyelin, into ceramides and free fatty acids. In particular acid and neutral sphingomyelinases generate ceramides with structural and signal transduction functions in epidermal proliferation and differentiation. The normal skin is colonized by a variety of microorganisms, both harmless commensals and potential pathogens, and effective defence mechanisms are essential for the protection of this barrier because, like the other human epithelia, it is under constant microbial assault [1, 2]. Antimicrobial peptides significantly contribute to the epithelial defence of multicellular organisms because they represent important effector molecules of the innate immune defence showing a broad-spectrum antimicrobial activity against a wide range of pathogens including bacteria, fungi and viruses. Additionally they play a crucial role as signalling molecules in linking innate and adaptive immune responses. Epidermal barrier function is abnormal in AD, with increased transepidermal water loss. Decreased stratum corneum ceramide content may cause the defect in permeability barrier function consistently found in both lesional and non-lesional skin of patients with AD. Jensen et al. [2] found decreased epidermal acid sphingomyelinase activity in lesional and non-lesional skin of patients with AD, correlating with reduced stratum corneum ceramide content and disturbed barrier function. This reduced ceramide level appears to be due to upregulation of sphingomyelin deacylase enzyme activity [3]. A defective permeability barrier leads to the penetration of environmental allergens into the skin and initiates immunological reactions and inflammation crucially involved in the pathogenesis of AD.

The stratum corneum of the skin of patients with AD is highly susceptible to colonization by various bacteria, mainly *S. aureus*. As one sphingolipid metabolite, sphingosine, is known to exert a potent antimicrobial effect on *S. aureus* at physiological levels, it may play a significant role in bacterial defence mechanisms of healthy normal skin. Because of the altered ceramide metabolism in AD, the possible alteration of the sphingosine metabolism might favour the vulnerability to colonization by *S. aureus* in patients with AD [3].

Dermatophagoides pteronyssinus is a trigger of AD. Many *D. pteronyssinus* allergens are proteases that can elicit airway inflammation by stimulating the release of cytokines and chemokines by bronchial epithelial cells [4]. *D. pteronyssinus* can cause proteolysis-dependent release of cytokines from keratinocytes, but it appears incapable of activating de novo expression of cytokines and chemokines, arguing against a direct pro-inflammatory activity of house dust mite (HDM) on the skin [4, 5].

The Skin of Atopic Patients: The Interaction with *S. aureus*

The skin of patients with AD is frequently overinfected by bacterial agents, especially *S. aureus*. Increased numbers of *S. aureus* are found in about 90% of skin lesions [6] and such bacterial colonization exacerbates or maintains skin inflammation by secreting superantigens which activate T cells and macrophages. In addition, *S. aureus* produces enzymes which can directly damage the skin cells [7]. Scratching probably enhances the binding between *S. aureus* and epidermal cells by exposing the extracellular matrix adhesion molecules [8]. The degree of colonization is associated with the disease severity [9, 10]. For these reasons topical and general antibacterial drugs are often used to keep the skin of atopic patients under control. Also steroids and topical immunomodulators appear to be able to reduce the degree of bacterial colonization [11, 12]. The presence of increased bacterial colonization has been demonstrated in patients with AD [6]. Such colonization has been included in the group of trigger factors for eczema in AD. The presence of bacterial agents, especially *S. aureus*, in the lesional areas of AD has also been demonstrated when clinically the lesions did not show any sign of impetiginization and the entity of bacterial colonization, even in non-lesional skin of patients with AD, is significantly higher than that observed in other cutaneous diseases [13–15]. The possibility to detect *S. aureus* in the skin of normal subjects and of children with AD varies with the different methods used. Using a dry standard swab in a previous work we observed a positive culture in the antecubital area in 38% of children affected by AD [16], and this percentage is similar to that reported in the recent work by Patel et al. [17]. A higher percentage of *S. aureus* has been detected with the use of the scrub technique and, more especially, with the contact plate technique. The latter method evaluates a wider sample area with a consequently higher possibility to detect bacteria. As observed in a recent paper [18], *S. aureus* was isolated from the majority of the bacterial colonies in subjects with AD; a significant relationship was also found between the total number of colonies and the severity of the eczema, as well as between severity and the number of *S. aureus* colonies.

Textiles

Garments are in direct contact with the skin all day long, and for this reason it is important to carefully choose suitable fabrics for atopic subjects who have disrupted skin.

Owing to their natural properties fabrics produced from non-synthetic fibres are preferential. Natural plant fibres include cotton, linen, hemp and jute, and animal fibres include wool and silk. Apart from these there are also artificial fibres produced from natural fibres like viscose, whose functional properties are close to those of natural fibres. Synthetic fibres, on the other hand, are produced by chemical processing of petroleum. At the beginning of their fabrication, all fibres undergo different processes to improve their characteristics or for dyeing. Some of the colourings used are important in determining sensitization to the skin or worsening atopic skin lesions.

Wool Fibres

Wool fibres are frequently used in human clothes but are irritant in direct contact with the skin. The main constituent of wool is keratin, a protein with disulphide bridges that cross-link chains in the polymer. Wool is thus characterized by higher breaking strength and environmental factors [19]. Although biodeterioration of wool is caused by both bacteria and fungi, keratin is degraded to a higher degree by fungi (*Microsporum*, *Aspergillus* and *Penicillium*). Wool fibre has frequently been shown to be irritant to the skin of atopic patients and for this reason wool intolerance was included as a *minor criterion* in the diagnostic criteria of AD by Hanifin and Rajka in 1980 [20]. The intensity of itching from wool fibres in AD has also been seen to increase in relation to the type of fibre: in a study on 24 girls with AD and a history of irritation to wool, Bendsoe et al. [21] observed that greater itching was provoked on normal skin of the abdomen by a material with coarse wool fibres (36 μm) than with thinner fibres (20 μm). Also in a recent paper by Williams et al. [22] about the factors influencing AD obtained by a questionnaire survey of schoolchildren's perceptions, wool fabrics together with sweating from exercise and hot weather were the three most common exacerbators, affecting 40, 41.8 and 39.1% of AD responders, respectively.

Cotton

Cotton is the most commonly used textile for patients with AD; it has wide acceptability as clothing material because of its natural abundance and inherent properties like good folding endurance, better conduction of heat, easy dyeability and excellent moisture absorption. In addition, its easy biodegradability is an added attraction. However, it suffers from drawbacks like inflammability, poor crease retention and is prone to bacterial and fungal attack. To modify the physical properties of fabrics (e.g. hydrophobicity, impermeability) and to improve their mechanical properties (softness, resilience, cohesion of the fibres), a coating treatment is carried out on woven or non-woven fabrics using an acrylic binder resin. Cotton is used for patients with AD despite its relative roughness: it is made up of many short fibres (1–3 cm) with flat and irregular sections; damp absorption and transfer occur by extension and contraction of the single fibres producing a movement that may sometimes irritate and scratch the skin. The irritation can also be related to the presence of detergents that act as a trigger factor. In a paper by Kiriyama et al. [23], the residual washing detergent in cotton clothes is demonstrated as being a factor of winter deterioration of dry skin in AD. The study followed 148 Japanese patients with AD during the winter months, who were instructed to wear cotton underwear washed with common detergents in cold tap water. The distribution of dry skin on their trunks was examined. They were then asked to stop washing their clothes with common anionic, additive-enriched detergents, and to use a non-ionic, additive-reduced detergent for a period of 2 weeks. The results suggested that residues of common washing detergents in cotton underclothes play an important role in the winter deterioration of dry skin in patients with AD who use cold tap water for washing their clothes. Cotton clothes specifically made for children with AD and used in conjunction with topical agents are available in some countries (e.g. Envicon® in Italy; Lohmann® in Germany). A recent study [24] has assessed the beneficial effects of softened fabrics on atopic skin, suggesting that a softened fabric is less aggressive to the skin than an unsoftened one. Furthermore, in the case of pre-irritated skin, its recovery was significantly faster when in contact with softened rather than unsoftened fabrics.

Reactive dyes are used especially for colouring natural fibres that are widely used in western countries, particularly Italy, in the production of clothes [25]. A recent study by Giusti et al. [26] has demonstrated that in children with suspected contact sensitization disperse dyes should be regarded as potential triggering allergens.

Silk

Silk in its natural state consists of a single thread secreted by the silkworm *Bombyx mori* and is made up of a double filament of protein material (fibroin) glued together with sericin, an allergenic gummy substance that is normally extracted during the processing of the silk threads [27, 28]. Sericin ensures the cohesion of the cocoon by gluing silk threads together. Most of the sericin must be removed during raw silk production. Degumming is the key process during which sericin is totally removed and silk fibres gain the typical shiny aspect, soft handle and elegant drape highly appreciated by consumers. Sericin is a water-soluble protein. Because of its properties, sericin is particularly useful for improving artificial polymers such as polyesters, polyamide, polyolefin and polyacrylonitrile. Sericin is also used as a coating or blending material for natural and artificial fibres, fabrics and polymer articles. The structure of silk fibre is quite similar to that of human hair (97% proteins, 3% fat and waxy substances), thus allowing its use in surgery and directly on scalded skin. Each silk thread is made up of many filaments more than 800 m long which are highly resistant, perfectly smooth and cylindrical and do not cause friction on the skin. Silk also helps to maintain the body temperature by reducing the excessive sweating and moisture loss that can worsen xerosis. Silk allergy among workers in the silk industry has been recognized for some time, whereas allergic reactions of consumers on a large scale have only been rarely described [29, 30] since the final textile products of silk are mostly non-allergenic [31]. After the removal of sericin, the remaining protein, fibroin, has a low allergenic potential, and only 1 case of recurrent granulomas with remarkable infiltration of eosinophils has been reported in the literature which may have resulted from an IgE-mediated hypersensitivity reaction to silk fibroin [32]. Although clothes made of washed silk are generally safe for the consumer, the unprocessed silk threads that are broken during reeling (silk waste) are used in China to fill bed quilts, clothes, toys and mattresses [31] and may continue to be allergenic. A study performed by Sugihara et al. [33] in Japan examined the effects of a silk film on full-thickness skin wounds: it was found that the wounds dressed with silk film healed 7 days faster than those covered with traditional dressing. They reported that silk film as a skin dressing can be easily produced and sterilized, and also enhances collagen synthesis, reduces oedema and scarring due to inflammatory responses and promotes epithelialization. Moreover, silk has been used as suture thread for many years, especially in epidermal and ophthalmic surgery [34]. However, the type of silk fabric generally used for clothes is not particularly useful in the care and dressing of children with AD since it reduces transpiration and may cause discomfort when in direct contact with the skin. A new type of silk fabric made of transpiring and slightly elastic woven silk is now commercially available (Microair Dermasilk®) and may be used for the skin care of children with AD. This fabric also has antibacterial properties thanks to an exclusive water-resistant treatment with Aegis AEM 5772/5, a durable antimicrobial finish for textile products that prevents odour and bacteria survival, including *S. aureus* [35]. It is based on the compound alkoxysilane quaternary ammonium. These Aegis antibacterial treatments are already utilized in the USA in many commercial products. Our group has tried to evaluate the efficacy of such fabrics. Forty-six children affected by AD in an acute phase were recruited: 31 received special silk clothes which they were instructed to wear for a week; the other 15 served as a control group and wore cotton clothing. Topical moisturizing creams or emulsions were the only topical treatment prescribed in both groups. The overall severity of the disease was evaluated using the SCORAD index. In addition, the local score of an area covered by the silk clothes was compared with the local score of an uncovered area in the same child. All patients were evaluated at baseline and 7 days after the initial examination. At the end of the study, a significant decrease in AD severity was observed in the children of group A (mean SCORAD decrease from 43 to 30; $p = 0.003$). At the same time, the improvement in the mean local score of the covered area (from 32 to 18.6; $p = 0.001$) was significantly higher than that of the uncovered area (from 31 to 26; $p = 0.112$). In a subsequent work we tried to evaluate in vivo the antibacterial activity of this special fabric with Aegis AEM 5572/5 that has shown in vitro antibacterial activity. Twelve children affected by AD with symmetric eczematous lesions on the antecubital areas and 4 without any cutaneous diseases as controls were recruited. Children had to wear the tubular dressing for 7 days, changing it every day. Microbiological examinations were obtained with standard cultural swab and by means of quantification of bacterial agents using agar plates at baseline, after 1 h and after 7 days. Preliminary data show a significant improvement in the mean value of the topical clinical SCORAD index in both the covered areas compared to the values obtained at baseline. Also the reduction of the mean number of colony-forming units per square centimetre was similar in both areas. As observed by other authors we confirm that the reduction of bacterial colonization in our cases is associated with a clinical improvement of eczema [36]. However, the preliminary analysis of our data shows that even though the silk fabric treated with Aegis shows in vitro antibacterial properties [37] we were unable to demonstrate such an activity in vivo. A possible explanation may be the difficulty to maintain a complete adherence between the fabric and the skin. The antibacterial efficacy of Aegis in vivo therefore remains not sufficiently demonstrated and the most realistic explanation is that the strong adhesion of Aegis to the fibre does not allow Aegis to penetrate into the skin and to contact bacteria sufficiently to kill them; the inability to reach bacteria in the lower layer of the stratum corneum or follicles would therefore allow them to survive. Furthermore, the rapid reinfection of eczema in AD by *S. aureus* leads one to suppose that in AD some bacteria may survive any type of antimicrobial treatment. The improvement observed also in the control areas was due to the textile protection bringing about a reduction in the external provocation factors associated with a better care of the eczema as frequently happens during a trial, and – already observed by Gauger et al. [36] – the advantage of the strong adhesion of Aegis to the fibre is that the possible toxicological side-effects of coated ammonium may be excluded because the ammonium is covalently bound to the textile fibre so closely that there is no absorption through the skin lesions.

Synthetic Fibres

Polyesters are high-molecular-weight compounds with repeated ester bonds in the main chain. The basic type of polyester fibre produced on a large scale is polyethylene terephthalate. Chemically it is a linear, saturated polyester of terephthalic acid and ethylene glycol (or ethylene oxide), and is also known as linear aromatic polyester. It has different trade names, e.g. Terylene (UK) and Dacron (USA), and is used for the production of fabrics, knitted fabrics, sheer curtains and technical fabrics. The macromolecules of the high-molecular-weight compounds are known as polyamides as they contain amide groups. The most important in the textile industry are the aliphatic polyamides (Steelon,

Perlon, Nylon, etc.). In the textile industry two kinds of polyurethane fibre are used: highly crystalline fibres of linear structure and highly flexible segment fibres of the Spandex type. Since they are characterized by high stiffness these polyurethanes are used in the form of monofilaments for brushes or as yarn for insulation material. The second type of polyurethane includes at least 85% urethane polymer of segmented structure and is used for flexible fibres. These polyurethanes are characterized by a lengthy elongation before breaking occurs in tests, and have many other advantages, including stable colour and high resistance to radiation and ageing. They are used in the textile industry in the production of fabrics, knitted fabrics, stockings, socks and bathing wear. The properties of polypropylene, produced by polymerizing propylene, depend, among other factors, on molecular weight, crystallinity and production methods. Polypropylene fibres are characterized by good mechanical properties and are readily used in the textile industry. Diepgen et al. [37] carried out a randomized clinical study on 55 patients with AD and 31 healthy controls to investigate the irritative capacity of poncho-like shirts made of 4 different materials (A: cotton; B, C, D: synthetics of different fibre structure). The intensity of itching or discomfort due to repeated wearing of these shirts was evaluated by means of a point system. The study clearly showed that the irritative capacity of synthetic shirts is significantly higher in patients with AD, while cotton shirts were best tolerated.

Other Textiles

Topical antibiotics or antiseptics, as well as the use of systemic antibiotics, are often necessary to reduce the aggressive action of bacterial agents in AD. Recently the antibacterial proprieties of a micromesh material containing woven silver filaments with a total silver content of 20% has been reported. Silver products have recently been demonstrated to offer two advantages in the control of bacterial infections: a broad-spectrum antimicrobial activity and the absence of drug resistance. Silver-coated materials are, in fact, already used in surgery. A textile consisting of micromesh material containing woven silver filaments (Padycare® textile) has been made and used in an open-labelled controlled side-to-side comparative trial: 15 patients diagnosed as having generalized or localized AD were evaluated regarding *S. aureus* colonization and clinical severity of AD over a 2-week period. Flexures of the elbows were covered with silver-coated textiles on one arm and cotton on the other for 7 days followed by a 7-day control period. A highly significant decrease in *S. aureus* colonization could be seen on the site covered by the silver-coated textile already 2 days after initiation and lasting until the end of the treatment. Seven days after cessation, *S. aureus* density remained significantly lower compared to baseline. In addition, significantly lower numbers of *S. aureus* were observed on the silver-coated textile site in comparison to cotton at the end of treatment as well as at the time of control. Clinical improvement was seen to correlate with the reduction of *S. aureus* colonization. However, a clinical improvement was also observed on day 2 on the cotton-treated site, probably due to the protective effect of the textile.

AD and Textiles as Prevention from Inhalant Allergens by Bedding Encasement

Textiles may be used not only for clothes but also to prevent dust mite sensitization in atopic patients. Several studies have evaluated the influence of the HDM on clinical symptoms in AD [38–40]. A marked clinical improvement of AD was observed by Tan et al. [41] in a group of adults and children with positive skin tests (not necessarily towards mites), after an intensive eradication programme for mite allergens. In an interesting double-blind placebo-controlled study, they observed significant benefits in a group of patients, predominantly adults, with AD who embarked on intensive dust avoidance measures (mattress encasing, high-filtration vacuum cleaning, use of acaricide and allergen-denaturing spray). The authors found that the greatest benefit from HDM avoidance seemed to occur in children [41]. There have been many attempts to eradicate HDM in the home by using bedding encasement, acaricides and preparations that denature antigens [42]. The use of encasing has been reported to be the most effective means of reducing the levels of HDM antigens [43], as well as being the most easily applied [44]. Following this observation, and as AD is most prevalent in the first years of life, we analysed the effects of dust avoidance with a special bedding encasement in a group of children. We enrolled patients significantly younger than those in the previous study but excluded children younger than 2 years to avoid the risk of food allergy interference. The patients were randomly assigned to an active mite avoidance group or to a control group at the beginning of the study. No significant differences in clinical severity or Der p 1 + Der f 1 concentration were observed between the two groups at the start of the study, while, by chance, dust load and consequently the allergen concentration were significantly different [41]. In the initial active mite avoidance group we observed an improvement in clinical score values already at the end of the first month; these stabilized after the second month, and further improvements occurred after 12 months. The significant decrease in dust load (from 371 to 204 mg m^{-2}) and concentration (from 731 to 299 ng m^{-2}) observed in the control group after the first month was particularly striking, even if a similar effect was observed by Tan et al. [41]. It is possible that our presence in the home of the patients to collect mite samples may have been a trigger to improve

HDM clearance. This hypothesis is partially confirmed by the observation that dust load (from 204 to 209 mg m^{-2}) and concentration (from 299 to 306 ng m^{-2}) values between the first and the second month were virtually unchanged in the control group, while in the mite avoidance group the concentrations were significantly lower in the second month (from 199 to 98 mg m^{-2} and from 230 to 94 ng m^{-2}) compared with the first month. We were aware that the avoidance measures proposed in this study were not particularly stringent, in that we chose to suggest only environmental measures with bedding encasing without the use of acaricides and allergen denaturants. There is still no agreement in the literature about the use of acaricides and allergen denaturants and their real efficacy, and some authors believe that they only have a limited role in controlling mite allergens [45]. On the other hand, the effectiveness of controlling mite allergens in beds by using bedding encasement with mite-blocking fibre is now well established [43, 45]. Another consideration influenced our choice of such avoidance measures: we wanted to suggest realistic measures that could be applied for a long period, even years, without our continuous supervision. We think that the slow decrease in HDM load may be related to the adoption of environmental avoidance measures alone. The data on HDM loads after 12 months, which show a decreased and similar value in both groups, support our initial approach. The decrease in Der p 1 + Der f 1 load from the initial values of 1.84 and 2.04 $\mu g\ g^{-1}$ of dust to the final values of 0.59 and 0.74 $\mu g\ g^{-1}$ of dust is similar to that reported by Nishioka et al. [43] in Japan, after 1 year of avoidance measures. The rapid clinical improvement observed in the initial active mite avoidance group may be related to the early decrease in HDM concentration: a small HDM reduction is probably sufficient to reduce skin hypersensitivity in patients with AD. In our opinion the most important parameter substantially changed by HDM avoidance measures, which could have brought about the clinical improvement in our patients, was the Der p 1 + Der f 1 concentration (micrograms per gram of dust). This parameter is also the one most frequently related to clinical improvement in asthma or other allergic respiratory diseases and also to mite sensitization [45]. This significant improvement over a brief period (1–2 months) occurred during the autumn and winter season, and was probably not associated with a spontaneous remission of the disease, as AD is usually exacerbated during these seasons [46]. On the other hand, we cannot exclude the possibility that the observed clinical improvement after 12 months of follow-up was partially linked to the spontaneous remission of AD in these children. In a previous study we detected HDM sensitization related to HDM concentrations in children with AD [47] in accordance with the results of other authors [43]. This sensitization may have already occurred by the first years of life [48], especially in children with associated food sensitization [49, 50]. HDM avoidance measures should be suggested to all families with children suffering from extrinsic AD, even in the first months of life, in order to control the cutaneous manifestations of AD and to prevent HDM sensitization. HDM sensitization in children with AD is the most important predictive factor for the appearance of allergic respiratory disease, above all asthma [51, 52]. Our data show that by using simple but efficacious environmental measures, a significant reduction in HDM in bedding takes place after 2 months. When avoidance measures are well explained, parents observing the improvement of AD in their children are motivated to maintain such a prevention for longer periods, as shown by the low HDM level observed at 12 months in both groups.

Other groups have evaluated the relevance of bedding encasement in AD: Oosting et al. [53] investigated, in a randomized, double-blind, placebo-controlled study, whether reducing HDM allergen levels by using mattress, duvet and pillow encasings for 12 months will result in improvement in AD symptoms. They found that use of HDM-impermeable encasings resulted in a significant decrease in Der p 1 and Der p 1 + Der f 1 allergen concentrations. However, this reduction in allergen load did not result in significant changes in clinical parameters between the groups. A recent study performed by Terreehorst et al. [54] studied the effects of bedding encasings in HDM allergy in 224 patients with asthma, rhinitis and AD; they demonstrated that bedding encasings do not improve quality of life in a mixed population of subjects with combinations with rhinitis, asthma and AD and sensitized to HDMs.

Textiles as Vehicle of Allergens

Clothing has been proposed as an additional source of exposure to mite and cat allergens. Dispersal of allergen into public places has also been attributed to clothing. In their paper De Lucca et al. [55] sought to study the contribution of various types of clothing on mite and cat exposure in a domestic environment. Also, they studied the ability of clothing to transfer allergens into a workplace. Personal exposure to mite and cat allergen from a range of clothing was measured by using intranasal air samplers in 11 homes. Five categories of clothing were tested. Wearing no upper clothing was the sixth category tested to distinguish the contribution of clothing from ambient background exposure. An adhesive tape was used to sample allergen from the surface of clothing, and reservoir dust samples were also collected. The above techniques were also used in the workplace to examine the amount of cat allergen transferred from cat owners to non-cat owners. The amount of mite and cat allergen inhaled differed among the clothing types worn and whether they had been washed recently. Wearing a woollen sweater increased personal allergen exposure to cat and mite allergen by a mean of 11 and 10 times, respectively. Clothing items that were less frequently washed carried more allergen whether assessed by vacuuming or sampled with adhesive tape. This corresponded to the amount of allergen inhaled. They also found that cat levels on non-cat owners' clothing increased significantly at the end of a working day, which led to an increase in their personal allergen exposure to cat dander. Another recent paper investigated the presence of HDMs on the skin and clothes of AD patients [56]. Skin surface samples were collected by sticky tape and examined by light microscopy. Dust samples from clothes and bedding were collected using a vacuum cleaner. The patients were examined before and after implementation of HDM environmental control measures including three skin treatments with an acaricide cream (5% permethrin). Mites were found on the skin of 8 out of 9 patients. Two patients were infested with *Dermatophagoides* spp. and the others with the hypopus stage of astigmatid mites. Two control patients were infested with *D. pteronyssinus* and *D. farinae*. Mites were found in all 43 dust samples. Mite numbers varied between 160 and 2,300 in clothing and between 30 and 5,100 per gram of dust in bedding. After implementation of mite control measures including acaricide skin treatment, the mean HDM score in bedding and clothing decreased from 43.6 ± 63.6 to 11.5 ± 11.2, and from 12.3 ± 12.5 to 7.7 ± 6.4 respectively; the mean clinical SCORAD decreased from 54.0 ± 20.4 to 21.7 ± 13.2 ($p = 0.001$). The HDM presence on the skin, clothing and bedding of AD patients may contribute to itching and transepidermal sensitization to HDM. The presence of mites in all developmental stages on human skin warrants further investigation of HDM biology and the human-mite relationship. Skin treatment with acaricide and HDM control measures can decrease AD symptoms.

Conclusion

An important role in the pathogenesis of AD has been assigned to the defect of the skin barrier which presents an impaired permeability function causing the development of eczematous lesions after exposure to repeated irritants [7]. The hypersensitivity of atopic skin may be improved or worsened by the use of textiles. The itching produced by direct contact with wool in patients with AD is characteristic and the irritation is likely to be caused by the 'spiky' nature of the fibres. In contrast to the worsening of the disease due to aggressive fabrics, the use of a fabric in the treatment of AD, especially of a soft one (cotton currently being the most utilized), reduces the contact irritation and prevents external bacterial agents [6]. However, recent studies have suggested that cotton may also present a roughness that irritates the skin of children affected by AD [19]. We nevertheless believe that the availability of soft fabrics with an adjunctive activity of antibacterial properties could be a new weapon without side-effects, and other studies need to be performed on such items. Clothes made of woven silk have special properties: the fabric allows the skin to breathe and the sensation does not bother the wearer; it also has a high capacity to absorb sweat and serous exudates (up to 30% of its weight without becoming damp): this is important in maintaining the water balance of the skin through its emollient and soothing action. The use of this new type of woven silk fabric would therefore appear to be useful in the skin care of children with AD. However, it should be stressed that the silk dressing is only effective if the garments are worn all day long, ensuring that contact with the skin is as close as possible. It is not easy for parents to follow this regimen strictly. Different textiles have various potential worsening links with allergies: e.g. clothing has been proposed as an additional source of exposure to mite and cat allergens. On the other hand, special textiles can be used to prevent dust mite sensitization. An ideal tissue should be soft and fresh, comfortable, easy to wash and dry, possibly composed of natural fibres and light colours, without disperse dyes. This ideal tissue should not catch dust or fur (especially that of cats); it should be without synthetic labels, and without seams, buttons or nickel elements that could worsen atopic skin irritation.

References

1 Taïeb A: Hypothesis: from epidermal barrier dysfunction to atopic disorders. Contact Dermatitis 1999;41:177–180.

2 Jensen JM, Folster-Holst R, Baranowsky A, Schunck M, Winoto-Morbach S, Neumann C, Schutze S, Proksch E: Impaired sphingomyelinase activity and epidermal differentiation in atopic dermatitis. J Invest Dermatol 2004;122:1423–1431.

3 Arikawa J, Ishibashi M, Kawashima M, Takagi Y, Ichikawa Y, Imokawa G: Decreased levels of sphingosine, a natural antimicrobial agent, may be associated with vulnerability of the stratum corneum from patients with atopic dermatitis to colonization by *Staphylococcus aureus*. J Invest Dermatol 2002;119:433–439.

4 Mascia F, Mariani V, Giannetti A, Girolomoni G, Pastore S: House dust mite allergen exerts no direct proinflammatory effects on human keratinocytes. J Allergy Clin Immunol 2002;109: 532–538.

5 Rieg S, Garbe C, Sauer B, Kalbacher H, Schittek B: Dermcidin is constitutively produced by eccrine sweat glands and is not induced in epidermal cells under inflammatory skin conditions. Br J Dermatol 2004;151:534–539.

6 Leung DYM, Bieber T: Atopic dermatitis. Lancet 2003;361:151–160.

7 Mempel M, Schnopp C, Hojka M, Fesq H, Weidinger S, Schaller M, Korting HC, Ring J, Abeck D: Invasion of human keratinocytes by *Staphylococcus aureus* and intracellular bacterial persistence represent haemolysin-independent virulence mechanisms that are followed by features of necrotic and apoptotic keratinocyte cell death. Br J Dermatol 2002;146:943–951.

8 Cho SH, Strickland I, Boguniewicz M, Leung DY: Fibronectin and fibrinogen contribute to the enhanced binding of *Staphylococcus aureus* to atopic skin. J Allergy Clin Immunol 2001;108: 269–274.

9 Bunikowski R, Mielke ME, Skarabis H, Worm M, Anagnostopoulos I, Kolde G, Wahn U, Renz H: Evidence for a disease-promoting effect of *Staphylococcus-aureus*-derived exotoxins in atopic dermatitis. J Allergy Clin Immunol 2000;105:814–819.

10 Tokura Y, Yagi J, O'Malley M, Lewis JM, Takigawa M, Edelson RL, Tigelaar RE: Superantigenic staphylococcal exotoxins induce T cell proliferation in the presence of Langerhans cells or class-II-bearing keratinocytes and stimulate keratinocytes to produce T-cell-activating cytokines. J Invest Dermatol 1994;102:31–38.

11 Brockow K, Grabenhorst P, Abeck D, Traupe B, Ring J, Hoppe U, Wolf F: Effect of gentian violet, corticosteroid and tar preparations in *Staphylococcus-aureus*-colonized atopic eczema. Dermatology 1999;199:231–236.

12 Remitz A, Kyllonen H, Granlung H, Reitamo S: Tacrolimus ointment reduces staphylococcal colonization of atopic dermatitis lesions. J Allergy Clin Immunol 2001;107:196–197.

13 Aly R, Maibach HI, Shinefield HR: Microbial flora of atopic dermatitis. Arch Dermatol 1977;113:780–782.

14 Leyden JJ, Marples RR, Kligman AM: *Staphylococcus aureus* in the lesions of atopic dermatitis. Br J Dermatol 1977;113:780–782.

15 Akiyama H, Ueda M, Toi Y, et al: Comparison of the severity of atopic dermatitis lesions and the density of *Staphylococcus aureus* on the lesions after antistaphylococcal treatment. J Infect Chemother 1996;2:70–74.

16 Ricci G, Patrizi A, Neri I, Bendandi B, Masi M: Frequency and clinical role of *Staphylococcus aureus* over-infection in atopic dermatitis of children. Pediatr Dermatol 2003;20:389–392.

17 Patel GK, Wyatt H, Kubiak EM, Clark SM, Mills CM: *Staphylococcus aureus* colonization of children with atopic eczema and their parents. Acta Derm Venereol 2001;81:366–367.

18 Arima Y, Nakai Y, Hayakawa R, Nishino T: Antibacterial effect of β-thujaplicin on staplylococci isolated from atopic dermatitis: relationship between changes in the number of viable bacterial cells and clinical improvement in an eczematous lesion of atopic dermatitis. J Antimicrob Chemother 2003;51:113–122.

19 Lee LD, Fleming BC, Waitkus RF, Baden HP: Isolation of the polypeptide chains of prekeratin. Biochim Biophys Acta 1975;18:82–90.

20 Hanifin JM, Rajka G: Diagnostic features of atopic dermatitis. Acta Derm Venereol 1980;92:44–47.

21 Bendsoe N, Bjornberg A, Asnes H: Itching from wool fibres in atopic dermatitis. Contact Dermatitis 1987;17:21–22.

22 Williams JR, Burr ML, Williams HC: Factors influencing atopic dermatitis – A questionnaire survey of schoolchildren's perceptions. Br J Dermatol 2004;150:1154–1161.

23 Kiriyama T, Sugiura H, Uehara M: Residual washing detergent in cotton clothes: a factor of winter deterioration of dry skin in atopic dermatitis. J Dermatol 2003;30:708–712.

24 Hermanns JF, Goffin V, Arrese JE, Rodriguez C, Piérard GE: Beneficial effects of softened fabrics on atopic skin. Dermatology 2001;202:167–170.

25 Manzini BM, Motolese A, Conti A, Ferdani G, Seidenari S: Sensitization to reactive textile dyes in patients with contact dermatitis. Contact Dermatitis 1996;34:172–175.

26 Giusti F, Massone F, Bretoni L, Pellicani G, Seidenari S: Contact sensitisation to disperse dyes in children. Pediatr Dermatol 2003;20:393–397.

27 Harindranath N, Prakash OM, Rao PVS: Prevalence of occupational asthma in silk filatures. Ann Allergy 1985;55:511–514.

28 Uragoda CG, Wijekoon PN: Asthma in silk workers. J Soc Occup Med 1991;41:140–142.

29 Borelli S, Stern A, Wüthrich B: A silk cardigan inducing asthma. Allergy 1999;54:892–902.

30 Celedon JC, Palmer LJ, Xu X, Wang B, Fang Z, Weiss ST: Sensitization to silk and childhood asthma in rural China. Pediatrics 2001;107:E80.

31 Chaoming W, Shitai Y, Lixin Z, Yu Y: Silk-induced asthma in children: a report of 64 cases. Ann Allergy 1990;64:375–378.

32 Kurosaki S, Otsuka H, Kunitomo M, Koyama M, Pawankar R, Matumoto K: Fibroin allergy: IgE mediated hypersensitivity to silk suture materials. Nippon Ika Daigaku Zasshi 1999;66:41–44.

33 Sugihara A, Sugiura K, Morita H, Ninagawa T, Tubouchi K, Tobe R, Izumiya M, Horio T, Abraham NG, Ikehara S: Promotive Effects of a silk film on epidermal recovery from full-thickness skin wounds. Exp Biol Med 2000;225:58–64.

34 Ratner D, Nelson BR, Johnson TM: Basic suture materials and suturing techniques. Semin Dermatol 1994;13:20–26.
35 Gettings RL, Triplett BL: A new durable antimicrobial finish for textiles. AATCC Book of Papers 1978, pp 259–261.
36 Gauger A, Mempel M, Schekatz A, Schafer T, Ring J, Abeck D: Silver-coated textiles reduce *Staphylococcus aureus* colonization in patients with atopic eczema. Dermatology 2003;207: 15–21.
37 Diepgen TL, Stabler A, Hornstein OP: Textile intolerance in atopic eczema – A controlled clinical study. Z Hautkr 1990;65:907–910.
38 Ring J, Darwson U, Gfesser M, Vieluf D: The atopy patch test in evaluating the role of aeroallergens in atopic dermatitis. Int Arch Allergy Immunol 1997;113:379–383.
39 Tupjer AR, De Monchy JGR, Coenraads PJ: Induction of atopic dermatitis by inhalation of house dust mite. J Allergy Clin Immunol 1996;97:1064–1070.
40 Sanda T, Yasue T: Effectiveness of house dust mite allergen avoidance through clean room therapy in patients with atopic dermatitis. J Allergy Clin Immunol 1991;89:653–657.
41 Tan BB, Weald D, Strickland I, Friedman PS: Double blind controlled trial of effect of house dust mite allergen avoidance on atopic dermatitis. Lancet 1996;347:15–18.
42 Warner JA, Marchant JL, Warner JO: Allergen avoidance in the house of asthmatic children. Clin Exp Allergy 1993;23:279–286.
43 Nishioka K, Yasueda H, Saito H: Preventive effect of bedding encasement with microfine fibers on mite sensitization. J Allergy Clin Immunol 1998;101:28–32.
44 Owen S, Morganstern M, Hepworth J, Woodcock A: A control of HDM antigen in bedding. Lancet 1990;335:396–397.
45 Platts-Mills TAE, Vervloet D, Thomas WR, Aalberse RC, Chapman MD: Indoor allergens and asthma: report of the Third International Workshop. J Allergy Clin Immunol 1997;100:S1–S24.
46 Werfel T, Kapp A: Environmental and other major provocation factors in atopic dermatitis. Allergy 1998;53:731–739.
47 Ricci G, Patrizi A, Specchia F, Menna L, Bottau P, D'Angelo V, Masi M: Mite allergen (Der p 1) levels in houses of children with atopic dermatitis: the relationship with allergometric tests. Br J Dermatol 1999;140:651–655.
48 Wahn U, Lau S, Bergmann R, Kulig M, Forster J, Bergmann K, Bauer CP, Guggenmoos-Holzmann I: Indoor allergen exposure is a risk factor for sensitization during the first three years of life. J Allergy Clin Immunol 1997;99:763–769.
49 Nickel R, Kulig M, Forster J, Bergmann R, Bauer CP, Lau S, Guggenmoos-Holzmann I, Wahn U: Sensitization to hen's egg at the age of twelve months is predictive for allergic sensitization to common indoor and outdoor allergens at the age of three years. J Allergy Clin Immunol 1997;99:613–617.
50 Sasai K, Furukawa S, Muto T, Baba M, Yabuta K, Fukuwatari Y: Early detection of specific IgE antibody against house dust mite in children at risk of allergic disease. J Pediatr 1996;128: 834–840.
51 Guillet G, Guillet MH: Natural history of sensitization in atopic dermatitis. Arch Dermatol 1992;128:187–192.
52 Ricci G, Patrizi A, Specchia F, Alboresi S, Arena G, Cannella MV, Di Lernia V, Masi M: Predictivity of prick test and RAST-positive reactions for inhaled allergens in children with atopic dermatitis for the appearance of allergic respiratory disease; in Czernielewski JM (ed): Immunological and Pharmacological Aspects of Atopic and Contact Eczema. Basel, Karger, 1991, vol 4, pp 177–178.
53 Oosting AJ, de Bruin-Weller MS, Terreehorst I, Tempels-Pavlica Z, Aalberse RC, de Monchy JG, van Wijk RG, Bruijnzeel-Koomen CA: Effect of mattress encasings on atopic dermatitis outcome measures in a double-blind, placebo-controlled study: the Dutch mite avoidance study. J Allergy Clin Immunol 2002;110:500–506.
54 Terreehorst I, Duivenvoorden HJ, Tempels-Pavlica Z, Oosting AJ, Monchy JG, Bruijnzeel-Koomen CA, Wijk RG: The effect of encasings on quality of life in adult house dust mite allergic patients with rhinitis, asthma and/or atopic dermatitis. Allergy 2005;60:888–893.

55 De Lucca SD, O'Meara TJ, Tovey ER: Exposure to mite and cat allergens on a range of clothing items at home and the transfer of cat allergen in the workplace. J Allergy Clin Immunol 2000;106:874–879.
56 Teplitsky V, Mumcouglu KY, Dalal I, Somekh E, Tanay A: House dust mites on the skin and clothes of atopic dermatitis patients. J Allergy Clin Immunol 2004;113:296.

Giampaolo Ricci, MD
Department of Paediatrics, University of Bologna
Via Massarenti 11
IT–40138 Bologna (Italy)
Tel. +39 051 6363639, Fax +39 051 636 4679, E-Mail ricci@med.unibo.it

Burg G (ed): Biofunctional Textiles and the Skin.
Curr Probl Dermatol. Basel, Karger, 2006, vol 33, pp 144–151

Coated Textiles in the Treatment of Atopic Dermatitis

S. Haug[a], *A. Roll*[a], *P. Schmid-Grendelmeier*[a], *P. Johansen*[a], *B. Wüthrich*[a], *T.M. Kündig*[a], *G. Senti*[a,b]

[a]Allergy Unit, Department of Dermatology, University Hospital of Zürich and
[b]Center for Medical Research, University Hospital of Zürich, Zürich, Switzerland

Abstract

Atopic dermatitis (AD) is a chronic inflammatory skin disease with increasing prevalence over the last few decades. Various factors are known to aggravate the disease. In particular, wool and synthetic fabrics with harsh textile fibres, aggressive detergents and climatic factors may exacerbate AD. Cutaneous superinfection, particularly with *Staphylococcus aureus*, is also recognized as an important factor in the elicitation and maintenance of skin inflammation and acute exacerbations of AD. The severity of AD correlates with *S. aureus* colonization of the skin. Beside the treatment of AD patients with creams and emollients, new developments in the textile industry may have therapeutic implications. Silk or silver-coated textiles show antimicrobial properties that can significantly reduce the burden of *S. aureus*, leading to a positive effect on AD. Silver products have been used as wound dressing, whereby silver has antiseptic properties, and drug resistance is hardly found. Padycare® textiles consist of micromesh material containing woven silver filaments with a total silver content of 20%. In vitro studies of these silver-coated textiles demonstrated a significant decrease in *S. aureus* and *Pseudomonas aeruginosa* as well as *Candida albicans*. Silk has been increasingly implemented in medical treatment of AD thanks to its unique smoothness that reduces irritation. Silk can be coated with antimicrobials (Dermasilk®). The combination of the smoothness of silk with an antimicrobial finish appears to make an ideal textile for patients suffering from AD.

Atopic Dermatitis and the Role of *Staphylococcus aureus*

Cutaneous superinfection, particularly with *Staphylococcus aureus*, is recognized as an important factor in the elicitation and the maintenance of skin inflammation and acute exacerbations of atopic dermatitis (AD) [1–9]. Hauser et al. [10] showed that, in comparison to healthy individuals, AD patients have significantly more *S. aureus* colonization, even in non-lesional skin. Moreover, in lesional skin of AD patients, *S. aureus* colonization is 100–1,000 times higher than in non-lesional skin [10], and the degree of colonization is associated with disease severity [11].

Knowledge of the pathophysiological role of *S. aureus* in AD has increased during recent years. The organism produces a variety of immunomodulatory toxins with superantigen properties such as the well-characterized staphylococcal enterotoxins A–E and the toxic shock syndrome toxin 1 [12]. In addition, *S. aureus* produces enzymes that directly exhibit cytotoxic properties, e.g. haemolysins and exfoliative toxins. *S. aureus* also produces elaborate defence mechanisms against the majority of currently used antimicrobial drugs [13, 14].

The immune response of AD patients to *S. aureus* is characterized by (1) a selective hyporesponsiveness to purified *S. aureus* cell walls when measured by delayed-type hypersensitivity in skin, (2) the induction of IgE antibodies against soluble and membrane-bound antigens of *S. aureus* in patients with high serum IgE titres and (3) regional lymphadenopathy that is not correlated with impetiginization but rather high total IgE and *S. aureus* cell-wall-specific IgE [15].

Although there is controversy about topical glucocorticoid and antibiotic combination therapy with regard to their potential diametrical pharmacological properties, antibiotics as well as antiseptic substances with antistaphylococcal activity are successfully used for the treatment of AD. Indeed, the reduction of the *S. aureus* burden adds to the anti-inflammatory effect of topical corticosteroids and emollients [16–18].

Atopic Dermatitis and the Role of Textiles

New developments in the textile industry may have secondary implications on medicine and may open new avenues in certain treatment traditions. For example, special silk or silver-coated textiles show antimicrobial properties (table 1). In addition to the absorbing and air-permeable natural substances such as cotton and wool, new materials have been developed. These new fabrics typically transmit humidity across the material and thereby help to keep the skin dry. This property may be beneficial in preventing fungal infections [19]. Likewise, textiles with antimicrobial coatings may be further means to prevent such infections. For instance, the admixture of chitosan to cotton fibres can be used to manufacture fabrics with antimicrobial properties.

Knittel et al. [20] published a report that describes new methods to modify the surface of textiles used near the skin. These modifications can readily be added to the conventional procedures of textile finishing, e.g. by using cyclodextrins or

Table 1. Silver-coated textile in comparison with silk fabric in the treatment of AD

Textile	Silver	Silk
Examples	Padycare®	Dermasilk®
Mode of action	Silver ions seem to cause a detachment of the cytoplasmic membrane from the bacterial cell wall; the existence of elements of silver and sulphur in the electron-dense granules and cytoplasm suggest the antibacterial mechanism of silver by loss of the ability of DNA replication and protein inactivation after Ag^{+} treatment	Silk properties thanks to an exclusive water-resistant treatment with Aegis ADM 5772/5, a durable antimicrobial finish for textile products that prevents odour and the survival of bacteria including *S. aureus*; it is based on the compound alkoxysilane quaternary ammonium → antibacterial (anti-germ)
Advantages	Broad-spectrum antibiotics, not yet associated with drug resistance (almost)	Allows the skin to breathe, high capacity to absorb sweat
Side-effects	–	–
Effects	Highly significant decrease in *S. aureus* colonization shown already after 2 days in children with AD	Dressing can be easily produced and sterilized, and also enhances collagen synthesis, reduces oedema and scarring due to inflammatory responses and promotes epithelialization

linear carbohydrate biopolymers that can be covalently attached to the textile, in order to allow frequent use and washing. Also Sander and Elsner [19] described the effect of antimicrobial textiles based on the admixture or treatment of the textile fibres with bactericidal chemicals or the subsequent applying of active substances. Textiles that have this antimicrobial finish prevent the bacterial metabolism of sweat components and can therefore prevent odour formation [20].

The defect of the skin barrier is assumed to play an important role in the pathogenesis of AD [21], as protection of the skin against exposure to irritants is reduced. Patients with AD often complain of itching when wearing woollen clothes. The itching is likely to be caused by the rather spiky nature of the wool fibres. Due to this phenomenon, parents of children suffering from AD are often advised to use cotton clothes for their children. However, recent studies have suggested that cotton may also irritate the skin of children affected by AD [22, 23].

Cotton is made up of many short fibres (1–3 cm) with flattened and irregular sections. Absorption and transfer of humidity occur by extension and contraction of the single fibres producing a movement that may irritate the skin. Novel textile materials can diminish this type of physical movement and thereby disrupt the itch-and-scratch cycles.

Silver

Since ancient times silver has been highly regarded as a versatile healing tool. In ancient Greece, Rome, Phoenicia and Macedonia, silver was used extensively to control infections and spoilage. Hippocrates taught that silver healed wounds and controlled disease. The popularity of medicinal silver especially arose throughout the mediaeval Middle East where it was widely used and esteemed for blood purification, heart conditions and to control halitosis. Paracelsus (approx. 1520) extensively used silver medicinally, and, following him, silver nitrate was successfully applied in the treatment of chorea and syphilis. In the 1800s, the antibacterial properties of silver were further described and clinically demonstrated. More recently, silver products have been investigated with special regard to wound-healing properties [24], whereby silver appears to have two key advantages: it is a broad-spectrum antibiotic [25], and drug resistance is hardly found [26]. Silver-coated materials are also routinely used in surgery (external fixation), urology (catheter) or odontology [27–29]. For the topical therapy of venous legs, Wunderlich and Orfanos [30] showed that a consistent therapy performed with dry wound dressings containing silver is superior to the conventional treatment without silver-containing dressings.

The antibacterial mechanism of action of silver is not yet fully understood, but silver ions seem to cause a detachment of the cytoplasmic membrane from the bacterial cell wall [31]. The existence of elements of silver and sulphur in the electron-dense granules and in the cytoplasm suggests that the antibacterial mechanism of silver may be impaired DNA replication and protein activation [31].

Padycare® textiles consist of micromesh material containing woven silver filaments with a total silver content of 20%. In vitro studies of these silver-coated textiles demonstrated a significant decrease in bacteria (*S. aureus* and *Pseudomonas aeruginosa*) as well as *Candida albicans* [24].

Gauger et al. [32] compared treatment with silver-coated textiles on one arm to that of cotton on the other arm for 7 days followed by 7 days without treatment in 15 patients with generalized or localized AD. This open-label controlled side-to-side comparative trial demonstrated a highly significant decrease in *S. aureus* colonization on the side covered by the silver-coated textile already after 2 days which lasted until the end of the treatment. Even 7 days after cessation, the *S. aureus* burden remained lower when compared to baseline. In addition, significantly lower numbers of *S. aureus* were observed on the surface of the silver-coated textile as compared to that of cotton. As the results of this study showed that clinical improvement was paralleled by reduced *S. aureus* colonization, this may point towards a crucial role of antiseptic therapy in the treatment of AD. These findings are in accordance with earlier studies implying that antibiotic or antiseptic therapy facilitates a faster clearance of AD [16, 33].

Another interesting finding of the study of Gauger et al. [32] was that the reduction of staphylococcal colonization was long lasting, with reduced bacterial burden more than 7 days after wearing the antimicrobial clothes. This is in contrast to the effects seen by the antistaphylococcal dye gentian violet where cessation of therapy resulted in immediate re-colonization by *S. aureus* [16]. This suggests that intermittent, e.g. overnight, wearing of silver-coated textiles might be sufficient to sustain impairment of *S. aureus* growth.

Finally, the toxicological side-effects of silver-coated textiles appear to be limited to systemic absorption through dermal wounds [34]. However, further studies on silver absorption in patients wearing silver-coated textiles need to be performed.

Silk

Silk in its natural state consists of a single thread secreted by the silkworm and is made up of a double filament of protein material (fibroin) glued together with sericin, an allergenic and gummy substance that is normally extracted during the processing of the silk threads [35, 36]. Silk is comprised of perfectly smooth fibres that do not cause mechanical irritation of the skin. The structure of silk fibres is quite similar to that of human hair (97% proteins, 3% fat and waxy substances), thus allowing its use in surgery and also directly on scalded skin. Each silk thread is made up of many filaments more than 800 m long which are highly resistant to mechanical and thermal forces. Silk helps to maintain the body temperature, by reducing the excessive sweating and moisture loss that can worsen xerosis. Whereas silk allergy among workers in the silk industry is widely recognized, allergic reactions of consumers on a large scale have been only rarely described [37, 38], as the final silk fabrics are mostly nonallergenic [39].

A study performed by Sugihara et al. [40] in Japan examined the effects of a silk film on full-thickness skin wounds. They found that wounds dressed with sterilized silk film healed 7 days faster than those covered with traditional dressing. The silk films also enhanced collagen synthesis, reduced oedema and scarring due to inflammatory responses and promoted epithelialization. Moreover, silk has been used as suture thread for many years especially in dermatological and ophthalmic surgery [41].

The type of silk fabric generally used for clothes is not particularly useful in the care and dressing of children with AD, as such silk reduces transpiration and may cause discomfort when in direct contact with the skin. However, a new type of silk fabric made of transpiring and slightly elastic woven silk is now commercially available (Microair Dermasilk®) and may be used for the skin care of children with AD. Woven silk allows the skin to breathe and the sensation is not skin irritating. It also has a high capacity to absorb sweat and serous exudates (up to 30% of its weight without becoming damp). The latter is important in maintaining the water balance of the skin through its emollient and soothing action. The use of sericin-free silk products would appear to alleviate the symptoms of AD in children and as such may represent a useful tool in the management of AD.

The Dermasilk also has antibacterial properties thanks to an exclusive water-resistant treatment with Aegis ADM 5772/5, a durable antimicrobial finish for textile products that prevents scent and the survival of bacteria including *S. aureus* [42]. It is based on the compound alkoxysilane quaternary ammonium. These Aegis antibacterial treatments are already utilized in the USA in many commercial products.

A study performed by Ricci et al. [43] examined the clinical effectiveness of a silk fabric in the treatment of AD. The study included 46 children aged between 4 months and 10 years with a mean age of 2 years. All children were affected by AD in accordance with the Hanifin and Rajka inclusion criteria [44]. Thirty-one children received products made of silk (Microair Dermasilk), a pure form of silk consisting exclusively of fibroin without sericin. Fifteen children in the control group received cotton clothing. No pharmacological treatment with steroids or antibiotics was permitted in either group. In addition, the local score of an area covered by the silk clothes was compared with the local score of an uncovered area in the same child. All patients were evaluated prior to and 7 days after the treatment start. At the end of the study, a significant decrease in AD severity was observed in the children wearing the silk clothes. At the same time, the improvement in the mean local score of the covered area was significantly greater than that of the uncovered area. While silk showed a statistically significant improvement of the skin, cotton showed no significant improvement; unpublished results from our group showed similar results. These data suggest that such special silk clothes may be useful in the management of AD in children.

Conclusion

The skin of children affected by AD is very sensitive and may worsen after exposure to various irritant factors. Such factors may include rough textile fibres, such as those in wool. Therefore cotton clothes have been recommended for children with AD. Also, a recent study demonstrated beneficial effects of softened fabrics on atopic skin [45], including a significantly faster recovery of irritated skin in contact with softened rather than unsoftened fabrics. Therefore, silk has been increasingly implemented in the medical treatment of AD thanks to its unique smoothness that reduces irritation.

The severity of AD correlates with *S. aureus* colonization of the skin. Silver-coated textiles induce a highly significant reduction of the *S. aureus* burden already within 2 days and show a positive clinical effect. These findings are in accordance with earlier studies implying that antibiotic or antiseptic therapy contributes to a faster clearance of AD [33].

Therefore, the combination of the smoothness of silk fabrics with an antimicrobial finish appears to make an ideal textile for patients suffering from AD.

References

1 Leung DY, Bieber T: Atopic dermatitis. Lancet 2003;361:151–160.
2 Abeck D, Bleck O, Ring J: Skin barrier and eczema; in Ring J, Behrendt H, Vieluf D (eds): New Trends in Allergy IV. Berlin, Springer, 1996, pp 213–220.
3 Imokawa G, et al: Decreased level of ceramides in stratum corneum of atopic dermatitis: an etiologic factor in atopic dry skin? J Invest Dermatol 1991;96:523–526.
4 Murata Y, et al: Abnormal expression of sphingomyelin acylase in atopic dermatitis: an etiologic factor for ceramide deficiency? J Invest Dermatol 1996;106:1242–1249.
5 Bleck O, et al: Two ceramide subfractions detectable in Cer(AS) position by HPTLC in skin surface lipids of non-lesional skin of atopic eczema. J Invest Dermatol 1999;113:894–900.
6 Werner Y, Lindberg M: Transepidermal water loss in dry and clinically normal skin in patients with atopic dermatitis. Acta Derm Venereol 1985;65:102–105.
7 Abeck D, Strom K: Optimal management of atopic dermatitis. Am J Clin Dermatol 2000;1:41–46.
8 Leung DY: Atopic dermatitis: the skin as a window into the pathogenesis of chronic allergic diseases. J Allergy Clin Immunol 1995;96:302–318, quiz 319.
9 Yarwood JM, Leung DY, Schlievert PM: Evidence for the involvement of bacterial superantigens in psoriasis, atopic dermatitis, and Kawasaki syndrome. FEMS Microbiol Lett 2000;192:1–7.
10 Hauser C, et al: *Staphylococcus aureus* skin colonization in atopic dermatitis patients. Dermatologica 1985;170:35–39.
11 Williams RE, et al: Assessment of a contact-plate sampling technique and subsequent quantitative bacterial studies in atopic dermatitis. Br J Dermatol 1990;123:493–501.
12 Tokura Y, et al: Superantigenic staphylococcal exotoxins induce T-cell proliferation in the presence of Langerhans cells or class II-bearing keratinocytes and stimulate keratinocytes to produce T-cell-activating cytokines. J Invest Dermatol 1994;102:31–38.
13 Mempel M, et al: Invasion of human keratinocytes by *Staphylococcus aureus* and intracellular bacterial persistence represent haemolysin-independent virulence mechanisms that are followed by features of necrotic and apoptotic keratinocyte cell death. Br J Dermatol 2002;146:943–951.
14 Chambers HC: Hackbarth, Methicillin-resistant staphylococci: genetics and mechanisms of resistance. Antimicrob Agents Chemother 1989;33:991–994.
15 Hauser C, et al: The immune response to *S. aureus* in atopic dermatitis. Acta Derm Venereol Suppl (Stockh) 1985;114:101–104.
16 Brockow K, et al: Effect of gentian violet, corticosteroid and tar preparations in *Staphylococcus-aureus*-colonized atopic eczema. Dermatology 1999;199:231–236.
17 Verbist L: The antimicrobial activity of fusidic acid. J Antimicrob Chemother 1990;25(suppl B):1–5.
18 Ring J, Brockow K, Abeck D: The therapeutic concept of 'patient management' in atopic eczema. Allergy 1996;51:206–215.
19 Sander C, Elsner P: Fungal infections and textiles. Akt Dermatol 2004;30:18–22.
20 Knittel D, et al: Functional textiles for skin care and as therapeutic medium. Akt Dermatol 2004;30:11–17.

21 Macheleidt O, Kaiser HW, Sandhoff K: Deficiency of epidermal protein-bound omega-hydroxy-ceramides in atopic dermatitis. J Invest Dermatol 2002;119:166–173.
22 Arcangeli F, Feliciangeli M, Pierleoni M: Indumenti di seta nella dermatite atopica. V Convegno Nazionale Dermatologia per il Pediatra, Bellaria, 2001, pp 100–101.
23 Bendsoe N, Bjornberg A, Asnes H: Itching from wool fibres in atopic dermatitis. Contact Dermatitis 1987;17:21–22.
24 Bioservice: Test 001118. 2001.
25 Lansdown A: Silver I: its antibacterial properties and mechanism. J Wound Care 2002;11:125–130.
26 Driver V: Silver dressings in clinical practice. Ostomy Wound Manage 2004;50(suppl 9A):11–15.
27 Bosetti M, et al: Silver coated materials for external fixation devices: in vitro biocompatibility and genotoxicity. Biomaterials 2002;23:887–892.
28 Schaeffer A, Story K, Johnson SM: Effect of silver oxide/trichloroisocyanuric acid antimicrobial urinary drainage system on catheter-associated bacteriuria. J Urol 1988;139:69–73.
29 Matsura T, et al: Prolonged antimicrobial effect of tissue conditioners containing silver zeolite. J Dent 1997;25:373–377.
30 Wunderlich U, Orfanos CE: Treatment of venous ulcera cruris with dry wound dressings: phase overlapping use of silver impregnated activated charcoal xerodressing. Hautarzt 1991;42:446–450.
31 Feng QL, et al: A mechanistic study of the antibacterial effect of silver ions on *Escherichia coli* and *Staphylococcus aureus*. J Biomed Mater Res 2000;52:662–668.
32 Gauger A, Mempel M, Schekatz A, Schäfer T, Ring J, Abeck D: Silver-coated textiles reduce *Staphylococcus aureus* colonization in patients with atopic eczema. Dermatology 2003;207:15–21.
33 Lever R, et al: Staphylococcal colonization in atopic dermatitis and the effect of topical mupirocin therapy. Br J Dermatol 1988;119:189–198.
34 Hollinger MA: Toxicological aspects of topical silver pharmaceuticals. Crit Rev Toxicol 1996;26:255–260.
35 Harindranath N, Prakash O, Subba Rao PV: Prevalence of occupational asthma in silk filatures. Ann Allergy 1985;55:511–515.
36 Uragoda CG, Wijekoon PN: Asthma in silk workers. J Soc Occup Med 1991;41:140–142.
37 Borelli S, Stern A, Wüthrich B: A silk cardigan inducing asthma. Allergy 1999;54:900–901.
38 Celedon JC, et al: Sensitization to silk and childhood asthma in rural China. Pediatrics 2001;107:E80.
39 Wen C, et al: Silk induced asthma in children: a report of 64 cases. Ann Allergy 1990;64:375–378.
40 Sugihara R, et al: Prevention of collagen-induced arthritis in DBA/1 mice by oral administration of AZ-9, a bacterial polysaccharide from *Klebsiella oxytoca*. Immunopharmacology 2000;49:325–333.
41 Ratner D, Nelson BR, Johnson TM: Basic suture materials and suturing techniques. Semin Dermatol 1994;13:20–26.
42 Gettings R, Triplett B: A new durable antimicrobial finish textiles. AATCC Book of Papers, 1978, pp 259–261.
43 Ricci G, Patrizi A, Bendandi B, Menna G, Varotti E, Masi M: Clinical effectiveness of a silk fabric in the treatment of atopic dermatitis. Br J Dermatol 2004;150:127–131.
44 Hanifin J, Rajka G: Diagnostic features of atopic dermatitis. Acta Derm Venereol (Stockh) 1980:92:44–47.
45 Hermanns JF, et al: Beneficial effects of softened fabrics on atopic skin. Dermatology 2001;202: 167–170.

Dr. Gabriela Senti
FMH Dermatology and Allergology
Clinical Trial Center, Center for Medical Research University Hospital of Zürich
Gloriastrasse 31
CH–8091 Zürich (Switzerland)
Tel. +41 44 255 1111, Fax +41 44 255 4418, E-Mail gabriela.senti@usz.ch

Burg G (ed): Biofunctional Textiles and the Skin.
Curr Probl Dermatol. Basel, Karger, 2006, vol 33, pp 152–164

Silver-Coated Textiles in the Therapy of Atopic Eczema

Anke Gauger

Klinik und Poliklinik für Dermatologie und Allergologie am Biederstein,
Technische Universität München, München, Deutschland

Abstract

Atopic skin is mainly determined by a disrupted skin barrier, resulting in a higher susceptibility to external irritants in affected and nonaffected skin. Apart from many other irritant and allergic influences, skin colonization with *Staphylococcus aureus* is one of the major factors triggering and maintaining atopic eczema (AE). Adequate textile protection with low irritant potential can be helpful in reducing the exposure to exogenous trigger factors. Until now, cotton fabrics have been the state of the art of recommended textiles for patients with AE. The combination of antimicrobial therapy with compatible textiles in terms of biofunctionality is a promising innovative approach. The antibacterial effect of silver-coated textiles on *S. aureus* colonization has been demonstrated in an open side-to-side comparison. Silver-coated textiles were able to reduce *S. aureus* density significantly after 2 days of wearing, lasting until the end of treatment (day 7) and even 1 week after removal of the textiles. In addition, there was a significant difference in *S. aureus* density comparing silver-coated with cotton textiles. In addition, the clinical efficacy and functionality of silver-coated textiles in generalized AE have been examined in a multicenter, double-blind, placebo-controlled trial. They were able to improve objective and subjective symptoms of AE significantly within 2 weeks, showing a good wearing comfort and functionality comparable to cotton without measurable side effects. These therapeutic effects led to a significantly lower impairment of quality of life, already after 2 weeks. Therefore, beside a potent antibacterial activity in vivo, silver-coated textiles demonstrate a high efficacy in reducing the clinical severity of AE showing a wearing comfort comparable to cotton.

Apart from many other influences, such as general and individual irritant and allergic factors, skin colonization with *Staphylococcus aureus* is known to play a major role in triggering and maintaining atopic eczema (AE).

The knowledge regarding the pathophysiological role of *S. aureus* in AE has increased over the last few years. Understanding the mechanisms underlying enhanced *S. aureus* colonization in AE and identification of the molecules involved in triggering skin inflammation have an important influence on the therapeutic approach to the disease [1–9].

Individual provocation factors play an important role in disease activity and therefore have to be diagnosed for each individual patient [7, 10]. Provocation factors also include nonspecific exogenous irritants affecting the skin barrier and leading to exacerbation. The disruption of the skin barrier function in patients with AE is known to be one of the major pathophysiological aspects of the disease [11]. Quantitative and qualitative changes in lipid composition [12–14] result not only in an increased transepidermal water loss [15], but also in a higher susceptibility to external irritants in affected and nonaffected skin [11]. Adequate textile protection can be helpful in reducing the exposure to exogenous trigger factors.

However, textiles themselves can act as powerful irritants, depending on their material and their texture. Therefore, adequate clothing with textiles of low irritant potential is essential for patients with AE. Until now, cotton fabrics have been the state of the art of recommended textiles for patients with AE, which are also frequently used in the therapeutic management of the disease [16]. Therefore, the demand of finding suitable textiles for patients with skin diseases, especially AE, is high.

The combination of antimicrobial therapy with compatible textiles in terms of biofunctionality is a promising innovative approach. In addition, one of the major aims in the therapy of a chronic disease is reduction of adverse effects.

Silver products have been under intensive investigation during recent years with special regard to wound healing. The antibacterial activity of silver is known as well as its low side effect potential [17–19]. Although the exact antimicrobial mechanisms of silver are not yet fully understood, silver products are already widely used in different medical fields. Silver-coated materials are already frequently used e.g. in wound care, surgery (external fixation), urology (catheter) or odontology [19–22]. Silver textiles offer new treatment modalities in AE with the two key advantages of showing broad-spectrum antimicrobial activity with negligible drug resistance [23]. The protective effects of comfortable clothes together with antibacterial effects may contribute to the clinical improvement of AE by eliminating important trigger factors.

Silver-Coated Textiles

Silver-coated textiles currently on the market (Padycare®) consist of micromesh material (82% polyamide, 18% Lycra) with woven silver filaments with a silver content of 20% in total. They are available in all sizes, for infants as well as adults (fig. 1). Textiles have to be worn tightly on the skin to ensure

Fig. 1. Silver-coated micromesh material is also available for children.

the interaction between fiber and skin. With these textiles, in vitro as well as in vivo studies have been carried out. In vivo, the antibacterial activity was investigated using a placebo-controlled side-to-side comparison [24]; for clinical efficacy, a double-blind, placebo-controlled study was carried out [25].

Antibacterial Activity

To a large extent the antibacterial activity of silver is still unclear. Investigations with $AgNO_3$ treatment and bacteria showed a detachment of the cytoplasmic membrane from the cell wall. A remarkable electron-light region appeared in the center of the cells, which contained condensed deoxyribonucleic acid (DNA) molecules. The existence of silver and sulfur elements in the electron-dense granules and cytoplasm detected by X-ray microanalysis suggested the antibacterial mechanism of silver: DNA lost its replication ability, and the protein became inactivated after Ag^+ treatment [17].

In vitro studies of the investigated silver-coated textiles demonstrate a significant decrease in bacteria (*S. aureus* and *Pseudomonas aeruginosa*) as well as *Candida albicans*. In contrast, the cytotoxic effect was found to be comparatively low [26].

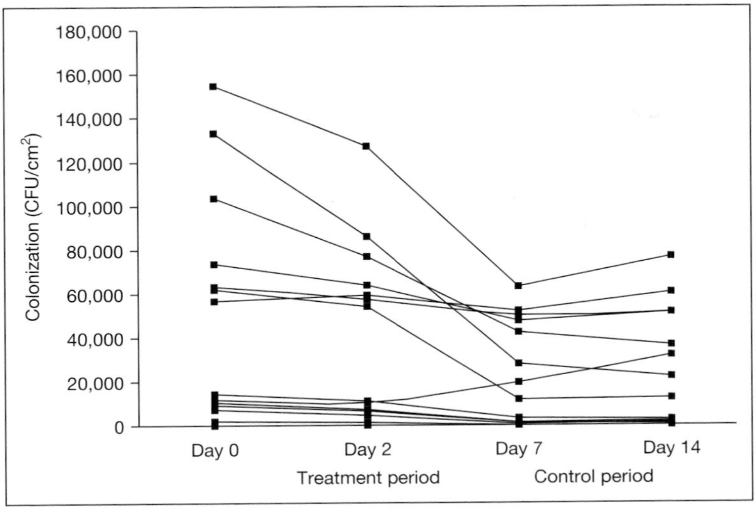

Fig. 2. Bacterial colonization by *S. aureus* in affected sites on the silver-coated textile site (right elbow flexure) at different time points of evaluation: days 0, 2, 7 and 14 in all 15 patients.

In a placebo-controlled side-to-side comparison, the antibacterial effect of silver-coated textiles on *S. aureus* colonization of affected sites was tested on the flexures of both elbows. In 15 patients diagnosed as having localized AE, these silver-coated textiles demonstrated a potent antibacterial activity which was accompanied by superior improvement in clinical severity of AE when compared to cotton [24]. Silver-coated textiles induced a highly significant reduction of *S. aureus* already 2 days after initiation of textile treatment and this lasted throughout the whole therapy phase (figs. 2, 3). Comparison between the silver-coated textile and cotton treatment sites revealed a significantly lower *S. aureus* colonization on day 7 ($p = 0.002$) and the time point of control (day 14; $p < 0.05$) on the silver-coated textile site. At baseline as well as on day 2, no significant difference between the two treatment modalities could be seen (fig. 4).

Even 7 days after cessation, *S. aureus* density was significantly lower when compared to baseline or cotton, showing a prolonged effect on staphylococcal reduction exhibiting the period of active wearing. This is in contrast to the effects seen by the antistaphylococcal dye gentian violet where cessation of therapy resulted in immediate subsequent recolonization [27]. These findings could indicate that overnight wearing of silver-coated textiles might be able to sustain a constant *S. aureus* reduction.

In the side-to-side-comparison, the reduction of *S. aureus* was paralleled by a reduction of clinical severity.

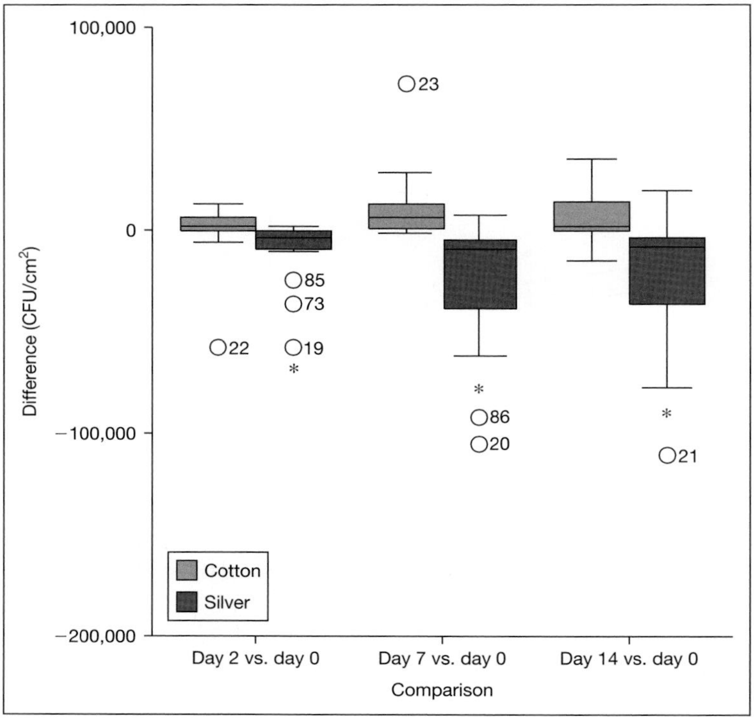

Fig. 3. Difference in *S. aureus* density within the affected silver-coated textile site (right elbow flexure) and cotton site (left elbow flexure) in 15 patients with AE during the study period (day 0 vs. days 2, 7 and 14). Reduction of *S. aureus* was highly significant on days 2, 7 and 14 on the silver-coated textile site (* $p < 0.01$) compared with baseline (day 0). On the cotton site, no significant reduction could be seen; ○ = extreme values beyond statistical analysis.

Clinical Efficacy

In the two studies carried out, an improvement in clinical severity could be noted in both, the placebo-controlled side-to-tide trial as well as the double-blind, placebo-controlled study.

SCORAD Assessment

The SCORAD ('Scoring Atopic Dermatitis') is a well-known and appropriate tool to measure the clinical severity of AE [28, 29]. This cumulative index combines objective (extent and intensity of different skin lesions) and subjective (daytime pruritus and sleep loss) criteria. Extent (A) of eczema lesions is assessed

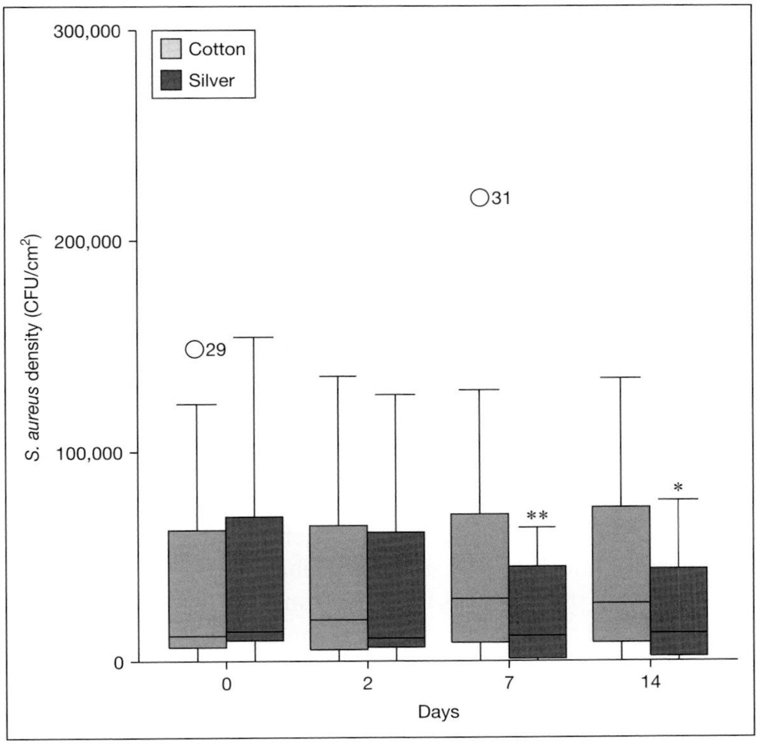

Fig. 4. Difference in *S. aureus* density between the silver-coated textile site (right elbow flexure) and cotton site (left elbow flexure) in 15 patients with AE during the study period (days 0, 2, 7 and 14). *S. aureus* colonization was significantly lower on days 7 and 14 on the silver-coated textile site (*$p < 0.05$, **$p < 0.01$) compared with cotton; ○ = extreme values beyond statistical analysis.

using the rule of 9, for intensity (B) 5 items are used: erythema (1), edema/papulation (2), oozing/crusts (3), excoriations (4), lichenification (5). Each item is graded on a 4-point scale: 0 = absent, 1 = mild, 2 = moderate, 3 = severe. Daytime pruritus and sleep loss (C) are evaluated by the patient using a visual analog scale from 0 to 10. The SCORAD is calculated by the following mathematical formula: SCORAD = $A/5 + 7B/2 + C$. Local disease severity can be assessed using the same 6 intensity items as in the general SCORAD. In addition, local pruritus is also measured on a 4-point scale. A total severity score as the sum of grading results in a highest possible score of 18.

The local and the general SCORAD are certified tools for the objective measurement of the clinical severity of AE. Silver-coated textiles were able to reduce the local as well as the general SCORAD in the two conducted placebo-controlled trials.

In the side-to-side comparison, reduction of *S. aureus* was paralleled by improvement in clinical severity. Already after 2 days, a constant decrease in the local SCORAD could be seen on the silver-coated textile site (right elbow flexure) in nearly all patients and continued to fall until termination of treatment (day 7). Thereafter, the severity of eczema was assessed as constant in half of the patients. The clinical improvement of local eczema was significant at all time points of evaluation (days 2, 7 and 14) in comparison with baseline ($p = 0.003$, $p = 0.001$ and $p = 0.004$, respectively).

In addition, when comparing the two treatment arms, the local SCORAD was constantly lower at all time points of clinical evaluation on the silver-coated textile site (right elbow flexure), reaching high statistical significance on day 7 and even 7 days after termination of treatment (day 14; fig. 5).

However, these promising results are of low expressiveness concerning the complex clinical picture of AE. Since AE is a systemic, not a local disease with a variety of individual and general influencing factors, the significance of silver-coated textiles on clinical severity including subjective symptoms as well as the irritative potential of these clothes is of major interest.

To investigate the aspects of general clinical improvement as well as the wearing comfort of these textiles, a double-blind, placebo-controlled study was carried out.

In this study, improvement of eczema was clearly shown by a significant reduction of the SCORAD. Reduction was significant only in the silver group after 7 and 14 days, whereas in the placebo group no statistical difference was noted between scores before and after the study (fig. 6). This may be explained by the antibacterial effect of silver leading to a reduction of the provocation factor *S. aureus*. However, there was no statistically significant difference between the two treatment groups. Clinical improvement was also seen in the placebo group, probably because of the additional treatment regimen with cotton textiles and the placebo effect of investigator and medical care.

Subjective Symptoms

Subjective symptoms, which are truly important in AE, were significantly reduced in the first week of treatment in the placebo (cotton) group (fig. 7). After 2 weeks, however, sleep loss and pruritus were also significantly diminished in the silver textile group (fig. 7). As recorded in the handed out questionnaire, significant differences between the silver group and the placebo group were observed after 7 days of wearing concerning the subjective assessment of pruritus improvement ($p = 0.02$). After 2 weeks, improvement of pruritus was still highly significant in the silver group when compared to cotton

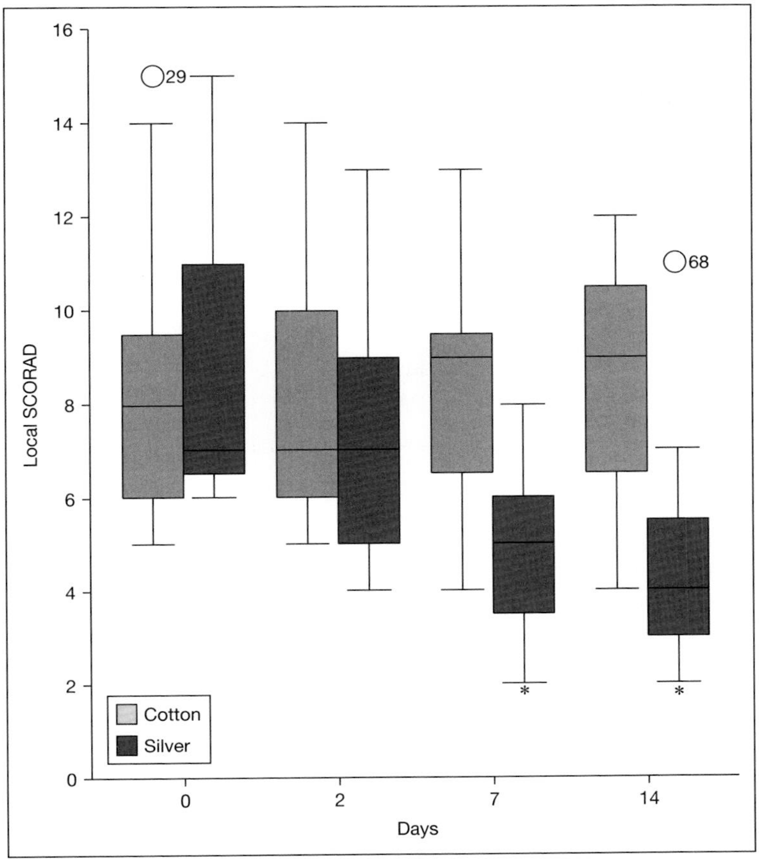

Fig. 5. Clinical eczema severity differences of the affected silver-coated textile site (right elbow flexure) compared with the cotton site (left elbow flexure) in 15 patients with AE. A significantly lower local SCORAD could be seen on days 7 and 14 ($*p < 0.01$) on the silver-coated site when compared with cotton. Clinical eczema severity is expressed in local SCORAD (0–18) in all 15 patients; ○ = extreme values beyond statistical analysis.

($p < 0.001$). Even the skin condition was assessed as being significantly better ($p = 0.003$) in the silver group and so was sleep improvement ($p = 0.02$), when evaluating the questionnaire.

In addition, silver textiles were able to reduce impairment in quality of life (QOL) significantly already after 2 weeks similar to cotton (fig. 8). QOL is an important factor regarding a chronic skin disease, and many studies have shown that all different aspects in QOL are affected by the disease [30]. The necessity to consider subjective symptoms is very high.

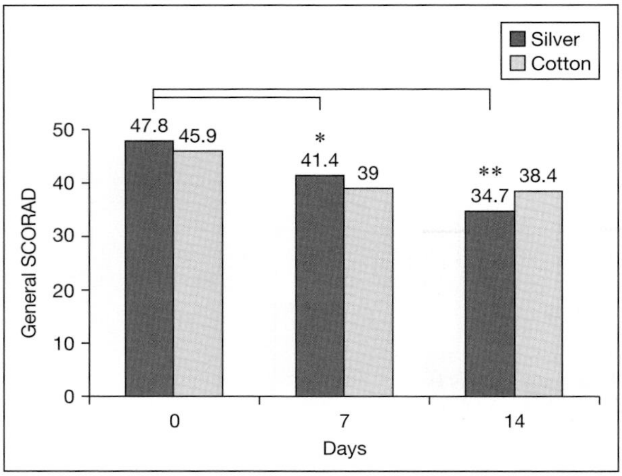

Fig. 6. General SCORAD (severity of eczema): significant improvement of general SCORAD after 1 week and after 2 weeks in the silver textile group. No statistical difference in the placebo group (cotton) between the condition before, during and after the study. $*p < 0.05$, $**p < 0.01$.

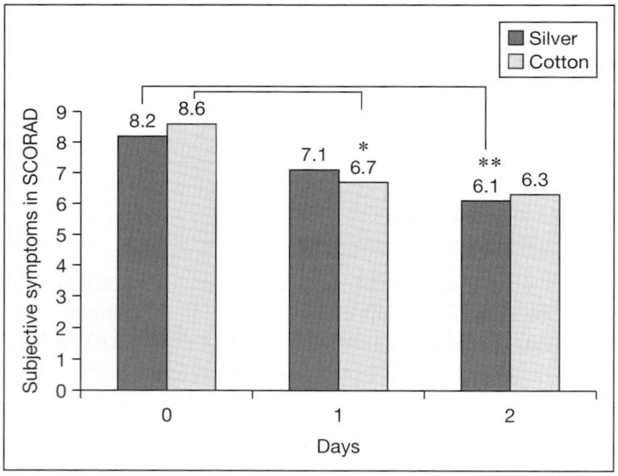

Fig. 7. Subjective symptoms in SCORAD evaluation (sleep loss and itching): significant reduction of subjective symptoms in the placebo group (cotton) in the first week, but not at the end of the study. In the silver textile group, significant reduction of sleep loss and itching after the study when compared with baseline. $*p < 0.05$, $**p < 0.01$.

Side Effects

During both studies, no side effects related to the study textiles were noted. Random blood samples taken from patients wearing the textiles beyond the

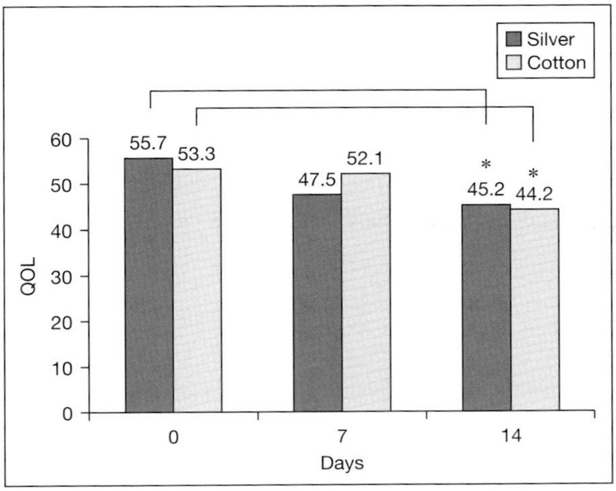

Fig. 8. Impairment of QOL: significant improvement at the end of the study in the cotton and silver textile groups. $^*p \leq 0.01$.

study conditions with their consent revealed no elevation of the serum silver level. However, a possible systemic absorption of silver because of a disrupted skin barrier has to be considered.

The use of topical steroids was documented in the placebo-controlled trial. More patients applied topical steroids in the placebo than in the silver group (84.4 vs. 68.6%, respectively) without statistical significance. At baseline, 42.9% of the silver group and 22.7% of the placebo group had been using topical steroids. After 2 weeks, steroid consumption was reduced to 24.6% in total, 28.6% in the silver and 18.2% in the placebo group. A tendency to a more pronounced difference in steroid use was noted in the silver group. However, no statistical difference between the two study groups or during the course of treatment was observed.

Irritative Potential

Irritative factors for AE include UV irradiation, water and detergents, sweat or extreme temperatures or temperature changes – from heat to cold or vice versa. As stated above, textiles themselves can bear an irritative potential on the disrupted skin barrier of AE patients. Therefore, textiles in the treatment of AE have to meet high standards to fulfill the patients' claims. Textiles have to be comfortable in all possible environmental conditions: cold, heat, dampness and dryness. They have to meet the patients' demands on hydrated or greasy as well as nonhydrated or dry skin. Until now, cotton garments have been the recommended textiles, since they demonstrate a low irritative potential. All-in-one suits made of cotton for infants are in regular therapeutic use [31].

In the study carried out, the wearing comfort of silver-coated textiles was comparable to that of cotton; no significant difference between the two treatment groups was noted. Furthermore, silver-coated study textiles have proven to show a better, but not statistically relevant, temperature regulation than cotton (73.5 and 50%, respectively) without relevant heat development (8.8 vs. 13.6%).

Conclusion

Taken together, the results of the studies carried out with silver-coated textiles for the treatment of AE further support the importance of their suitability. The clinical efficacy of silver-coated textiles together with a good antibacterial profile has been shown. In addition, a high functionality as well as a high wearing comfort of these textiles have been demonstrated.

Adequate clothes are known to contribute to barrier and eczema stabilization [11, 16, 31].

Beside multiple individual provocation factors, microbial stimuli such as bacteria or fungi, especially *S. aureus*, are recognized as important provocation factors of AE [1–7]. The degree of colonization has been found to be associated with disease severity [2, 4, 8]. Silver-coated textiles were able to significantly reduce *S. aureus* colonization together with a potent therapeutic improvement of eczema in objective and subjective parameters. These findings are in accordance with the clinical experience that antiseptic therapy is essential for an efficient therapy of affected lesions in AE [24] and earlier studies showing that antibiotic or antiseptic therapy is essential for a faster clearance of AE [27, 32].

A clinical effect of cotton textiles could also be demonstrated supporting the role of their suitability in the therapy of AE, i.e. their function to protect the skin from provocation factors. Beside irritative mechanisms, allergic reactions to aeroallergens are known to be major triggering factors, and exposure to relevant allergens may lead to exacerbation and/or maintenance of AE. The most important aeroallergens are house dust mites, pollen and animal dander, such as that of cats, dogs or horses. Beside allergen avoidance, which may be problematic in case of ubiquitous aeroallergens such as pollen, protection of the exposed skin from provocation factors by suitable textiles is the most important measure. In addition, textiles protect the inflamed or sensitive skin from scratching sequelae (disruption of the 'itch-and-scratch cycle') and can therefore be regarded as important therapeutic tools [16, 31].

In summary, it has been clearly shown that silver-coated textiles have a potent antibacterial activity in vivo and are able to reduce the clinical severity of

AE within a wearing period of 2 weeks significantly without side effects. Silver-coated textiles are comfortable and comparable to cotton concerning their wearing comfort and functionality, pruritus could be diminished even more effectively than with cotton. These therapeutic effects lead to a significantly lower impairment of QOL, already after 2 weeks. The use of silver-coated textiles may possibly lead to a less frequent use of topical corticosteroids or the use of less potent corticosteroids; however, until now there have been no extensive data available. In addition, the amount and the effect of silver ions detached from the textiles as well as possible resorption effects in patients wearing silver-coated textiles need to be further investigated.

References

1 Leung DYM: Role of *Staphylococcus aureus* in atopic dermatitis; in Bieber T, Leung DYM (eds): Atopic Dermatitis. New York, Dekker, 2002, pp 401–418.

2 Bunikowski R, Mielke ME, Skarabis H, Worm M, Anagnostopoulos I, Kolde G, Wahn U, Renz H: Evidence for a disease-promoting effect of *Staphylococcus-aureus*-derived exotoxins in atopic dermatitis. J Allergy Clin Immunol 2000;105:814–819.

3 Yarwood JM, Leung DY, Schlievert PM: Evidence for the involvement of bacterial superantigens in psoriasis, atopic dermatitis, and Kawasaki syndrome. FEMS Microbiol Lett 2000;192:1–7.

4 Nomura I, Tanaka K, Tomita H, Katsunuma T, Ohya Y, Ikeda N, Takeda T, Saito H, Akasawa A: Evaluation of the staphylococcal exotoxins and their specific IgE in childhood atopic dermatitis. J Allergy Clin Immunol 1999;104:441–446.

5 Mempel M, Schmidt T, Weidinger S, Schnopp C, Foster T, Ring J, Abeck D: Role of *Staphylococcus aureus* surface-associated proteins in the attachment to cultured HaCaT keratinocytes in a new adhesion assay. J Invest Dermatol 1998;111:452–456.

6 Leung DYM: Atopic dermatitis: the skin as a window into the pathogenesis of chronic allergic diseases. J Allergy Clin Immunol 1995;96:302–318.

7 Ring J, Abeck D, Neuber K: Atopic eczema: role of microorganisms on the skin surface. Allergy 1992;47:265–269.

8 Williams RE, Gibson AG, Aitchison TC, Lever R, Mackie RM: Assessment of a contact-plate sampling technique and subsequent quantitative bacterial studies in atopic dermatitis. Br J Dermatol 1990;123:493–501.

9 Bode U, Ring J, Neubert U: Intrakutantestungen und bakteriologische Untersuchungen bei Patienten mit atopischem Ekzem. Allergologie 1982;5:259–261.

10 Fischer S, Ring J, Abeck D: Atopic eczema: spectrum of provovocation factors and possibilities for their effective reduction and elimination. Hautarzt 2003;54:914–924.

11 Abeck D, Bleck O, Ring J: Skin barrier and eczema; in Ring J, Behrendt H, Vieluf D (eds): New Trends in Allergy IV. Berlin, Springer, 1996, pp 213–220.

12 Bleck O, Abeck D, Ring J, Hoppe U, Vietzke JP, Wolber R, Brandt O, Schreiner V: Two ceramide subfractions detectable in Cer(AS) position by HPTLC in skin surface lipids of nonlesional skin of atopic eczema. J Invest Dermatol 1999;113:894–900.

13 Imokawa G, Abe A, Jin K, Higaki Y, Kawashima M, Hidano A: Decreased level of ceramides in stratum corneum of atopic dermatitis: an etiologic factor in atopic dry skin? J Invest Dermatol 1991;96:523–526.

14 Murata Y, Ogata J, Higaki Y, Kawashima M, Yada Y, Higuchi K, Tsuchiya T, Kawainami S, Imokawa G: Abnormal expression of sphingomyelin acylase in atopic dermatitis: an etiologic factor for ceramide deficiency? J Invest Dermatol 1996;106:1242–1249.

15 Werner Y, Lindberg M: Transepidermal water loss in dry and clinically normal skin in patients with atopic dermatitis. Acta Derm Venereol 1985;65:102–105.

16 Ring J, Brockow K, Abeck D: The therapeutic concept of 'patient management' in atopic eczema. Allergy 1996;51:206–215.
17 Feng QL, Wu J, Chen GQ, Cui FZ, Kim TN, Kim JO: A mechanistic study of the antibacterial effect of silver ions on *Escherichia coli* and *Staphylococcus aureus*. J Biomed Mater Res 2000;52: 662–668.
18 Hollinger MA: Toxicological aspects of topical silver pharmaceuticals. Crit Rev Toxicol 1996;26:255–260.
19 Lansdown AB, Williams A, Chandler S, Benfield S : Silver absorption and antibacterial efficacy of silver dressings. J Wound Care 2005;14:155–160.
20 Bosetti M, Masse A, Tobin E, Cannas M: Silver coated materials for external fixation devices: in vitro biocompatibility and genotoxicity. Biomaterials 2002;23:887–892.
21 Schaeffer AJ, Story KO, Johnson SM: Effect of silver oxide/trichloroisocyanuric acid antimicrobial urinary drainage system on catheter-associated bacteriuria. J Urol 1988;139:69–73.
22 Matsura T, Abe Y, Sato Y, Okamoto K, Ueshige M, Akagawa Y: Prolonged antimicrobial effect of tissue conditioners containing silver zeolite. J Dent 1997;25:373–377.
23 Lansdown AB: Silver. I. Its antibacterial properties and mechanism. J Wound Care 2002;11: 125–130.
24 Gauger A, Mempel M, Schekatz A, Schäfer T, Ring J, Abeck D: Silver-coated textiles reduce *Staphylococcus aureus* colonization in patients with atopic eczema. Dermatology 2003;207:15–21.
25 Gauger A, Fischer S, Mempel M, Schäfer T, Weidinger S, Foelster-Holst R, Abeck D, Ring J: Efficacy and functionality of silver-coated textiles in patients with atopic eczema. J Eur Acad Dermatol Venereol, in press.
26 Bioservice: Test 001118. 2001.
27 Brockow K, Grabenhorst P, Abeck D, Traupe B, Ring J, Hoppe U, Wolf F: Effect of gentian violet, corticosteroid and tar preparations in *Staphylococcus-aureus*-colonized atopic eczema. Dermatology 1999;199:231–236.
28 Kunz B, Oranje AP, Labreze L, Stalder JF, Ring J, Taïeb A: Clinical validation and guidelines for the SCORAD index: consensus report of the European Task Force on Atopic Dermatitis. Dermatology 1997;195:10–11.
29 European Task Force on Atopic Dermatitis: Severity scoring of atopic dermatitis – The SCORAD index: consensus report of the European Task Force on Atopic Dermatitis. Dermatology 1993;186: 23–31.
30 Grob JJ, Revuz J, Ortonne JP, Auquier P, Lorette G: Comparative study of the impact of chronic urticaria, psoriasis and atopic dermatitis on the quality of life. Br J Dermatol 2005;152:289–295.
31 Abeck D, Strom K: Optimal management for atopic dermatitis. Am J Clin Dermatol 2000;1: 41–46.
32 Lever R, Hadley K, Downey D, Mackie R: Staphylococcal colonization in atopic dermatitis and the effect of topical mupiricin therapy. Br J Dermatol 1988;119:189–198.

Dr. Anke Gauger
Department of Dermatology and Allergy Biederstein, Technical University Munich
Biedersteinerstrasse 29
DE–80802 Munich (Germany)
Tel. +49 89 4140 3396, Fax +49 89 4140 3578, E-Mail agauger@gmx.de

Burg G (ed): Biofunctional Textiles and the Skin.
Curr Probl Dermatol. Basel, Karger, 2006, vol 33, pp 165–178

A New Silver-Loaded Cellulosic Fiber with Antifungal and Antibacterial Properties

Uta-Christina Hipler, Peter Elsner, Joachim W. Fluhr

Department of Dermatology and Allergology, Friedrich Schiller University of Jena, Jena, Germany

Abstract

The skin is the interface between the body and the environment. Each skin type has a specific skin physiology and is more or less adapted for protection against multiple stress factors. Textiles on the other hand are the tissues with the longest contact with the human skin. They play a critical role especially in skin conditions with an increased rate of bacterial and fungal infections like atopic dermatitis or hyperhidrosis, diabetic patients and aged skin. The present study demonstrates the antifungal and antibacterial effect of Sea Cell® Active in an in vitro test system against *Candida albicans* (DSM 11225), *Candida tropicalis* (ATCC 1169) and *Candida krusei* (ATCC 6258). Furthermore, the antibacterial activity of fibers with different amounts of Sea Cell Active fibers could be demonstrated in a dose-dependent manner against *Staphylococcus aureus* (ATCC 22923) and *Escherichia coli* (ATCC 35218). Whether this fiber seems to be suited for bioactive textiles in specific anatomical body regions and skin conditions with a susceptibility to fungal and bacterial infections, namely *Candida* species, *S. aureus* and *E. coli,* must be examined by means of further investigations, especially human in vivo tests considering allergic and toxic effects of the fiber.

The skin is the interface between the body and the environment. Each skin type has a specific skin physiology and is more or less adapted for protection against multiple stress factors [1–4]. Textiles on the other hand are the tissues with the longest contact with the human skin [5]. They play a critical role especially in skin conditions with an increased rate of bacterial and fungal infections like atopic dermatitis or hyperhidrosis, diabetic patients and aged skin [6–9].

The increasing demand for 'intelligent' and 'bioactive' textiles inspired the German company Alceru GmbH (Schwarza-Rudolstadt) to develop a new fiber

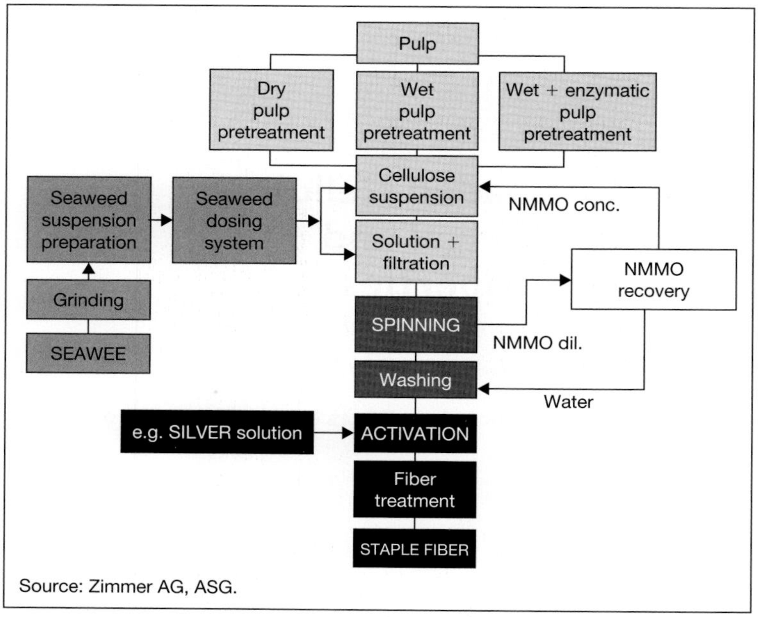

Fig. 1. Schematic overview of the Sea Cell Active production according to the Alceru process. NMMO = N-methylmorpholine-N-oxide.

called Sea Cell® Active [10]. The fiber can be manufactured by means of the so-called Lyocell® process (fig. 1). This process has been established as an environment-friendly, economically viable, product-enhancing and highly flexible alternative for the manufacture of man-made cellulose fibers. In the Lyocell process, cellulose is dissolved directly without formation of derivatives [11, 12]. The nontoxic, aqueous N-methylmorpholine-N-oxide is used as a solvent.

The spinning solution is processed in a combined dry/wet spinning step (air gap) to form fibers and shaped cellulose articles [13, 14]. During this spinning process, the solvent required to produce the spinning solution is washed out and almost completely recovered (fig. 1).

The Sea Cell fibers are manufactured by adding finely ground seaweed, mainly from the family of brown, red, green and blue algae. Particularly the brown algae *Ascophyllum nodosum* and/or the red algae *Lithothamnium calcareum* are added to form the spinning solution [15–19] (fig. 2).

The algae are added either as a powder or as a suspension in one of the process steps preceding the spinning of the cellulose solution. Seaweed has the capability of absorbing the minerals contained in seawater. An analysis showed

Fig. 2. A. nodosum (***a***), fiber, yarn and fabric (***b***).

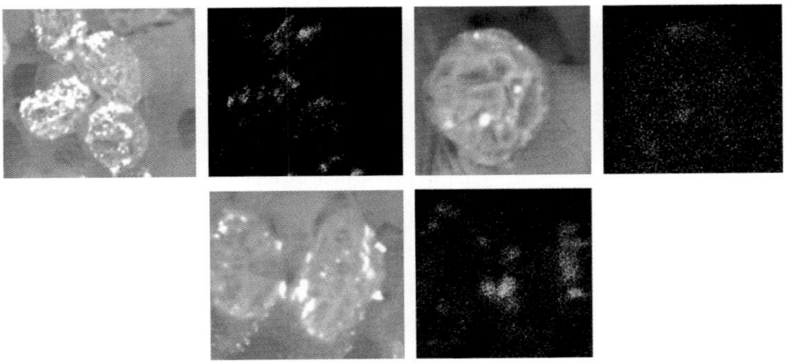

Method:
- Cryogenic break of the fibers
- Scanning electron microscope (LEO, type DSM 962) with
- EDX analysis (energy dispersive X-ray spectroscopy – Oxford Instruments)

Source: Zimmer AG, ICF Denkendorf.

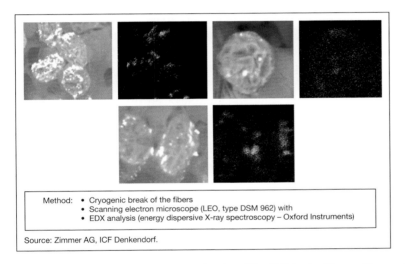

Fig. 3. Examination of cryogenic breaks of Sea Cell Active fibers with scanning electron microscopy and energy-dispersive X-ray analysis (distribution of silver at the break surface).

that in addition to minerals also carbohydrates, amino acids, fats and vitamins are detectable in seaweed.

Given this variety of active ingredients, seaweed and/or seaweed extracts are used in cosmetics as well as in the pharmaceutical industry [20–25].

In addition the Sea Cell fiber also exhibits a remarkably high tensile strength in dry and wet condition as well as negligible shrinking. Based on the good physical properties of the textiles, fabrics made from Sea Cell fibers offer high dimensional stability in addition to high wear comfort. One particularity of the Sea Cell fiber is its capacity to bind and absorb substances. During the activation of Sea Cell fibers, bactericidal metals like silver, zinc, copper and others can be absorbed by the fully formed cellulosic fiber through metal absorption. Unlike the commonly used method of incorporating the active ingredients in the spinning solution, the manufacture of Sea Cell Active offers the possibility of incorporating the substance permanently in the core of the fully formed fiber in an activation step. Impregnation tests with diluted metal salt solutions showed that the Sea Cell fiber exhibits excellent sorptive properties regarding metals and/or metal ions. It must be assumed that the metals are bound via free carbonyl, carboxy and hydroxyl groups of the cellulose as well as of the incorporated seaweed (fig. 3).

It is well known that phenols contained in the seaweed have the ability to chelate heavy metals [26, 27]. The metal ions are firmly anchored in the fiber matrix through the swelling of the cellulose which promotes an even distribution of the seaweed over the fiber cross-section. Even conventional cleaning methods that usually require an alkaline atmosphere do not affect the metal concentrations in the silver-loaded Sea Cell fiber. To obtain fibers with a permanent antibacterial finishing, ionic silver and zinc solutions are used for loading/activating.

The natural, cellulose and seaweed-based Sea Cell fibers serve as a functional carrier for the active compound silver, which has been known for more than a century to exert antifungal and antibacterial activity [28]. Additionally, the seaweed-based, silver-loaded Lyocell fiber Sea Cell Active contains the minerals calcium, magnesium and sodium, which are known to play a key role in skin homeostasis [29, 30].

The present study was intended to test whether Sea Cell Active with different amounts of the active silver-loaded fiber exerts antifungal and antibacterial properties. We were interested in testing the antifungal activity against several different fungi from the *Candida* family: *Candida albicans* (DSM 11225), *Candida tropicalis* (ATCC 1169) and *Candida krusei* (ATCC 6258). *C. albicans* is responsible for a widely encountered itching skin infection with yeasts especially in skin folds. These fungal infections are associated with warm, moist and occlusive conditions, e.g. under the armpits, under the breasts as well as in the genital and anal regions. Figure 4a and b gives some clinical examples of infections with *C. albicans* in the interdigital and genital (child) regions.

The second part of this study was intended to test whether Sea Cell Active with different amounts of the active silver fiber exerts antibacterial properties. The antibacterial activity against 2 different *Staphylococcus aureus* strains (ATCC 22926) and 1 *Escherichia coli* strain (ATCC 35218) should be tested. *S. aureus* is responsible for severe skin infection and the wide spread of multiresistant strains is becoming a major problem not only in dermatology.

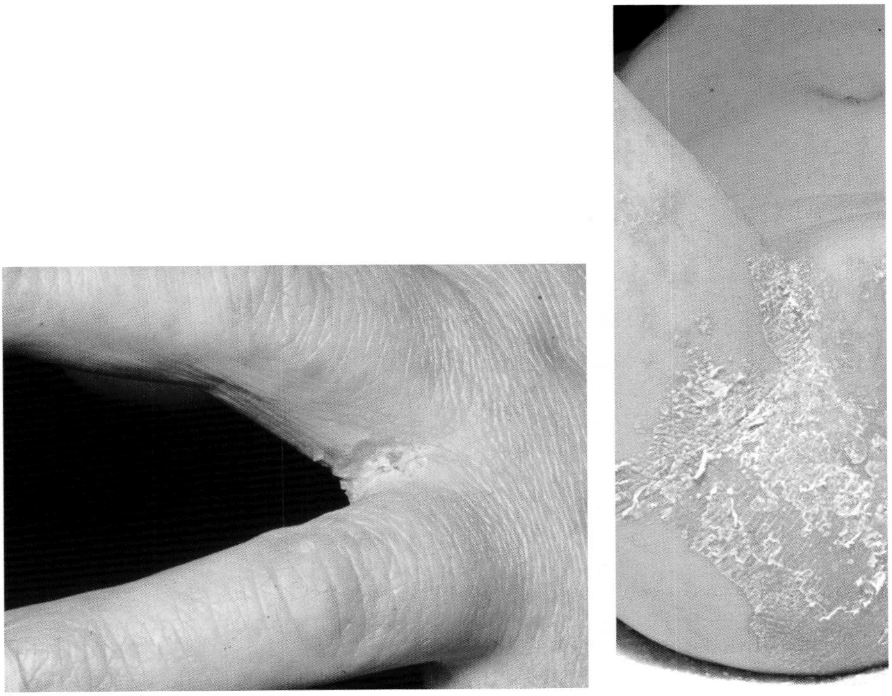

a *b*

Fig. 4. Dermal infection with *C. albicans* in the interdigital (***a***) and genital (***b***) regions.

Figure 5 shows a dermal infection with *S. aureus*. Furthermore *S. aureus* is known as an aggravation factor in atopic dermatitis. *E. coli* is a bacterium frequently encountered in interdigital and genital infections.

Materials and Methods

The following compositions were tested:
SCA 100: Sea Cell Active 100%;
SCA 100nw: Sea Cell Active 100% nonwoven;
SCA 5–20: Sea Cell Active 5–20%;
LC 100: Lyocell (control without silver loading);
SC 100: Sea Cell (control without silver loading);
w: washed fiber.

Two different in vitro test systems were used. First a classical incubation of standard strains with different fibers was performed over 24 h. The antifungal effect was quantified in a Neubauer cell-counting chamber. Additionally the fungi were stained with a fluorescent dye (FUN-1). FUN-1 belongs to a new family of fluorescent probes developed for assessing the

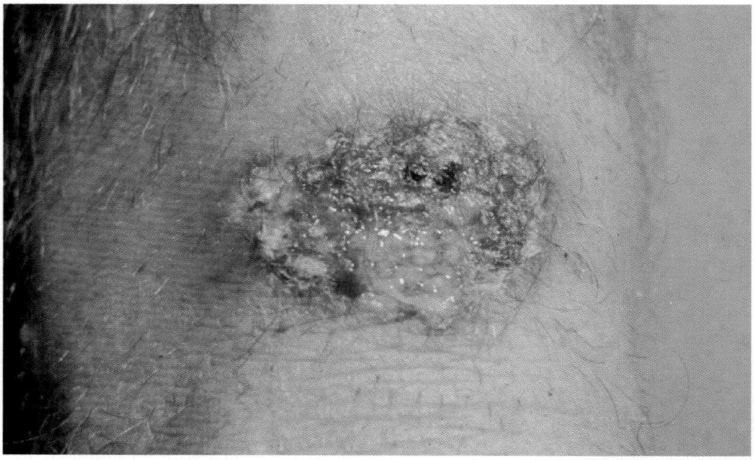

Fig. 5. Dermal bacterial infection with *S. aureus.*

viability and metabolic activity of fungi. FUN-1 [2-chloro-4-(2,3-dihydro-3-methyl-[benzo-1,3-thiazol-2-yl]-methylidene)-1-phenylquinolinium iodide] is a membrane-permanent nucleic-acid-binding dye that stains cylindrical intravacuolar structures in fungi [31]. Biochemical processing of the dye by active cells yielded cylindrical intravacuolar structures that were markedly red shifted in fluorescence emission and therefore spectrally distinct from the nucleic-acid-bound form of the dye. Overnight cell cultures of the *Candida* species were prepared. Subsequently, cells were counted and adjusted using cell counting with CASY® 1 to a final working concentration between 3×10^5 and 3×10^6 cells/ml in 20 ml Sabouraud glucose bouillon. This number of cells was inoculated with 40 mg test materials. The fibers were ground to a powder by means of a ball grinder prior to incubation. These cultures were incubated at 30°C for 24 h. Then 200 μl from each overnight cell culture was suspended with 1 ml glucoheptonate solution. The suspension was centrifuged at 14,000 g for 5 min. The pellets obtained were resuspended again in 50 μl glucoheptonate solution. From the last suspension 10 μl was added to 10 μl of FUN-1 (10 μ*M*). After incubation at 30°C for 30 min, 10 μl of cell suspension was trapped between a microscope slide for microscopy. Microscopy was carried out with an Olympus fluorescence microscope BX 40 equipped with ×20, ×40 and ×100 objective lenses. Epifluorescence illumination was provided by a 50- or a 100-watt mercury arc lamp. An excitation filter of 480 nm and an emission filter ≥530 nm were used. Photomicrographs were acquired with an Olympus digital camera BX 40 + C 5050 zoom. Photographic slides were digitalized electronically, and composite figures were assembled from the resulting images with Analysis® (Soft Imaging System GmbH, Münster, Germany).

The inhibition zone was determined for antibacterial activity [32]. Colonies of a pure culture of test bacterium were inoculated in 5 ml Mueller-Hinton broth and incubated at 37°C until the turbidity of the suspension was equal to that of 0.5 McFarland.

The cellulosic fiber materials were cut into 6-mm diameter disks. These disks were then placed in the center of Mueller-Hinton agar plates that had been inoculated with test bacteria at 37°C for 24 h. Thereafter the zones of inhibition were measured.

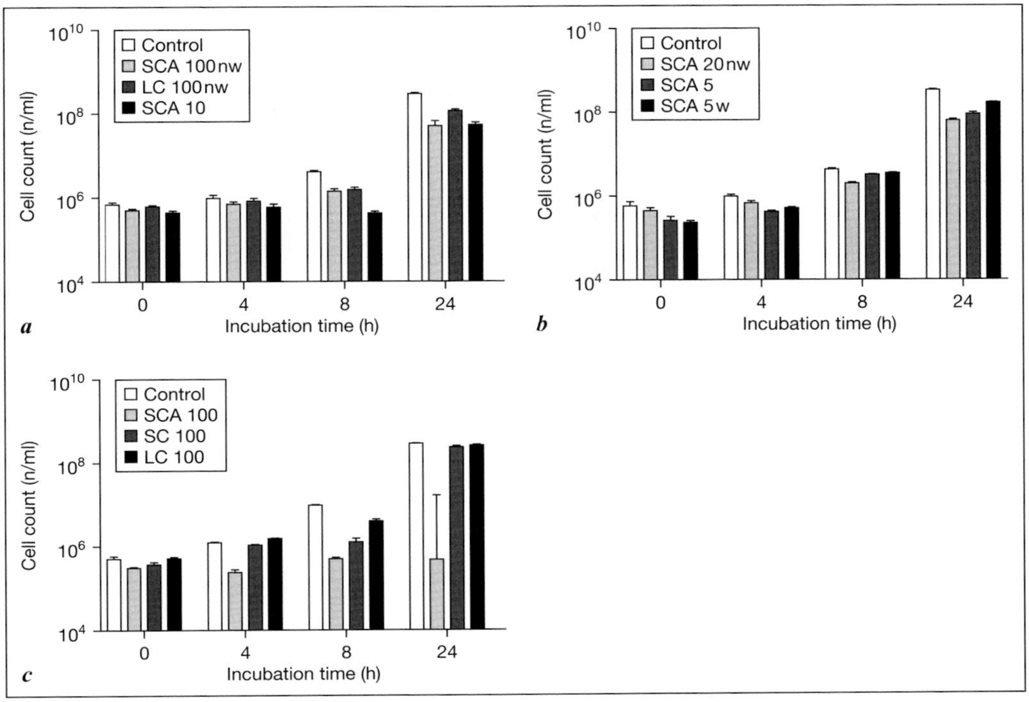

Fig. 6a–c. Antifungal activity against *C. albicans* (strain DSM 11225) of different fibers in a dose-dependent manner. The highest antifungal activity was detected for Sea Cell Active 100% (SCA 100). Measurement values are calculated on the basis of 8 single readings (means ± SD).

Statistics

Each measurement was performed 8 times. Means and standard deviations were calculated by means of the Microsoft® Excel software and Prein 3.02 (Graphpad, San Diego, Calif., USA).

Results

The results of the present study revealed an excellent antifungal activity against different fungi from the *Candida* family, namely *C. albicans* (fig. 6a–c), *C. krusei* (fig. 7a–c) and *C. tropicalis* (fig. 8a, b). Other *Candida* species like *C. parapsilosis* (onychomycosis) and *C. glabrata* (genital infection) were also susceptible to the Sea Cell Active fibers (data not shown). The most striking result was the dose-dependent manner (percentage of Sea Cell Active fibers in the tissue) of the antifungal activity against all. The highest antifungal activity was

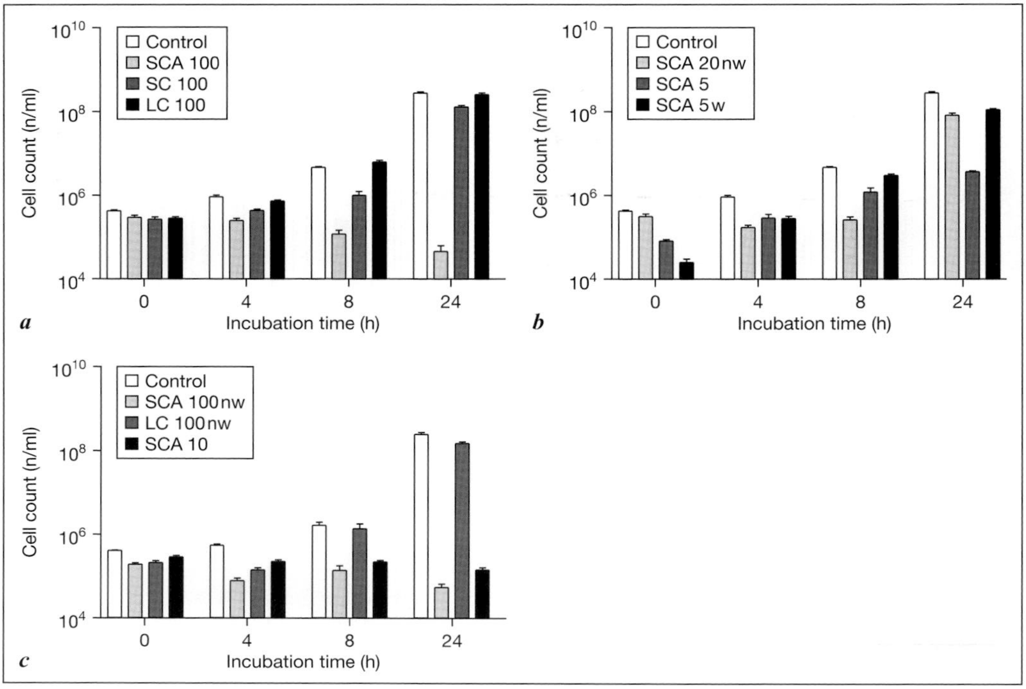

Fig. 7a–c. Antifungal activity against *C. krusei* (strain ATCC 6258) of different fibers in a dose-dependent manner. The highest antifungal activity was detected for Sea Cell Active 100% (SCA 100 and SCA 100nw). Measurement values are calculated on the basis of 8 single readings (means ± SD).

found for Sea Cell Active 100% (SCA) in all 3 *Candida* species. After 24 h of incubation, the cell counts only amounted to 10–20% compared with the control. The activity of Sea Cell Active 100nw is slightly smaller. These results were confirmed by FUN-1 staining (fig. 9) where the difference between untreated, active fungi and after treatment with Sea Cell Active 100% can be observed. Almost all living cells of *C. albicans* and *C. krusei* will be killed by the Sea Cell Active fibers.

In a second test series, the activity of the silver-covered seaweed-based fiber against the growth of the bacteria strains *S. aureus* (fig. 10a, b) and *E. coli* (fig. 11a, b) could be shown. This antibacterial activity was dose dependent with the highest activity in 100% Sea Cell Active fibers with 100% of the active silver load and lowest in the Lyocell and Sea Cell. An intermediate activity was detected in a mixed tissue with 5% of Sea Cell Active and more pronounced with 10 and 20% of the silver-loaded fiber.

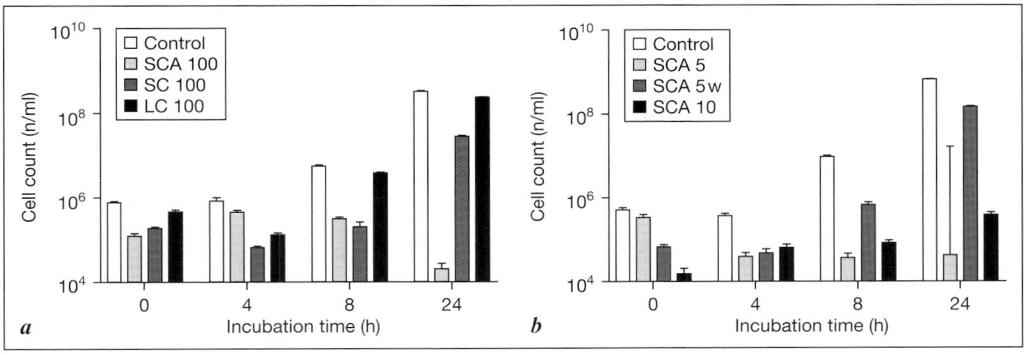

Fig. 8a, b. Antifungal activity against *C. tropicalis* (strain ATCC 1169) of different fibers. The highest antifungal activity was detected for Sea Cell Active 100% (SCA 100). Measurement values are calculated on the basis of 8 single readings (means ± SD).

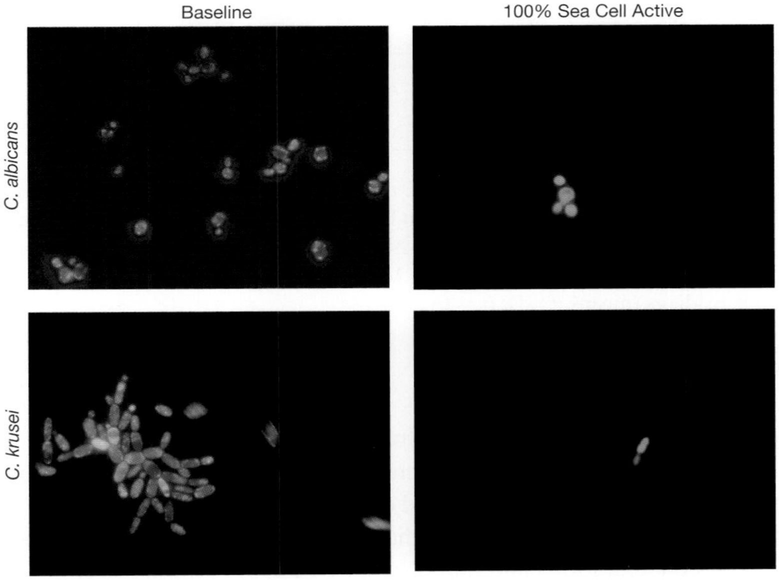

Fig. 9. Staining with fluorescent dyes (FUN-1): baseline staining showed biochemically active fungi while 100% Sea Cell Active treatment killed almost all living *C. albicans* (strain DSM 11225) and *C. krusei* (strain ATCC 6258).

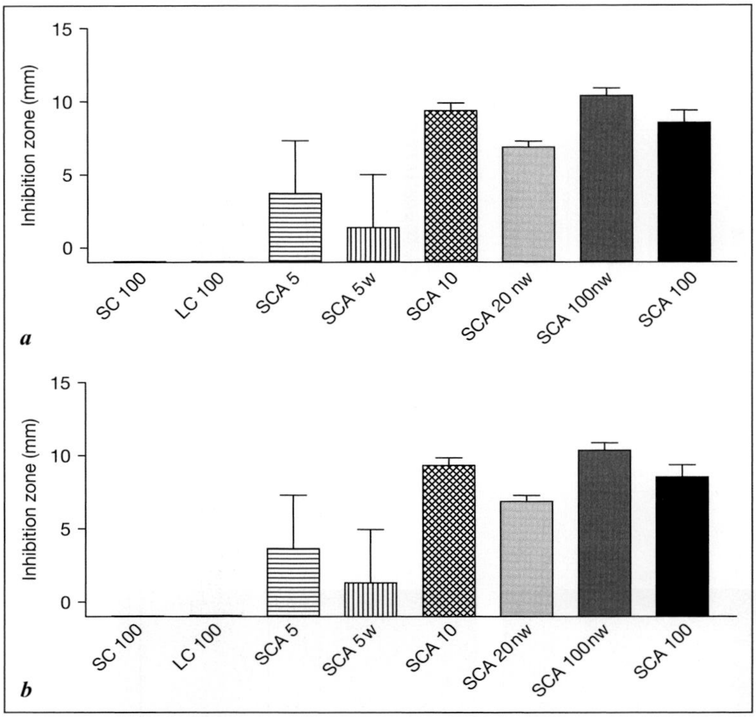

Fig. 10a, b. Antibacterial activity against *S. aureus* (strain ATCC 22923) of different fibers in a dose-dependent manner at two different bacterial concentrations in the inoculum: incubation 10^6 CFU/ml (***a***) or 10^5 CFU/ml (***b***) for 24 h. The highest antibacterial activity was detected for Sea Cell Active 100% (SCA 100 nw). Measurement values are calculated on the basis of 8 single readings (means ± SD; n = 6).

Discussions

The antibacterial effect of silver was already known in ancient times. Silver tools and containers (approx. 4000 BC) were used for storing and transporting water to prevent the formation of germs and ensure high water quality [33].

In the 19th century, it was evidenced that silver has an antimicrobial effect even in smallest concentrations and that it quickly destroys typhoid bacilli and the resistant anthrax spores at concentrations as low as 1:4,000 to 1:10,000 [28, 33]. In the decontamination of water as well as in the disinfection of wounds, the oligodynamic effect of silver (inhibition of bacterial growth already at metal concentrations of 0.006–0.5 ppm) is used [33, 34]. Bedding and textiles with minor concentrations of silver already show positive effects in the treatment of allergic dermatitis or psoriasis [35].

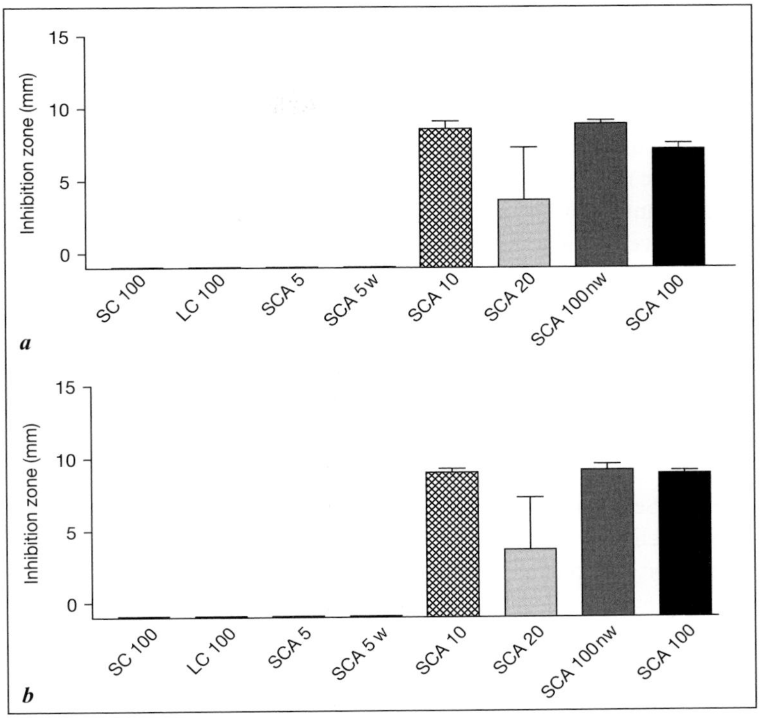

Fig. 11a, b. Antibacterial activity against *E. coli* (strain ATCC 35218) of different fibers in a dose-dependent manner at two different bacterial concentrations in the inoculum: incubation 10^6 CFU/ml (***a***) or 10^5 CFU/ml (***b***) for 24 h. The highest antibacterial activity was detected for Sea Cell Active 100% (SCA 100nw). Measurement values are calculated on the basis of 8 single readings (means ± SD; n = 6).

Silver nitrate and silver sulfadiazine have been used as prophylaxis or adjunctive therapy for sepsis in the care of burn wounds since the 1960s [28, 36]. Studies have revealed that silver sulfadiazine dissociates reacting with cellular DNA and that only silver ions bind DNA and inhibit its replication [37, 38]. Silver ions also interfere with the respiratory chain at the cytochromes and in the reduced nicotinamide adenine dinucleotide-succinate dehydrogenase region [39].

However, in silver products the silver release is very quickly inactivated by chloride and proteins on the skin surface or in wound exudates [40]. Therefore frequent reapplication of the agent is required. As a consequence Sea Cell Active was developed to provide a controlled, sustained release of silver. Due to the very low but sufficient leaching out of silver ions from the Sea Cell Active fiber, the antibacterial activity remains unchanged over time as shown for the washed fiber (approx. 60 wash cycles). As silver does not have any negative side effects like skin irritation, this metal is favored for use as an activating agent to manufacture Sea Cell Active fibers [41]. From a dermatological point of view it is therefore ensured that no skin irritation occurs. The excellent wear comfort of the cellulosic fiber is not affected.

This study revealed the excellent antifungal and antibacterial properties of Sea Cell Active in vitro. The results are in good agreement with the findings of Yin et al. [42] investigating the antimicrobial activity of Acticoat, a barrier dressing, in comparison with other conventional topical antimicrobial agents. These results demonstrate that the silver extracted from the dressing has the lowest minimum inhibitory concentrations and minimum bacterial concentrations. However, if silver contents in the 2 other silver-based compounds – silver nitrate and silver sulfadiazine – are calculated, almost the same minimum inhibitory and minimum bacterial concentrations are observed for all 3 silver products. This supports the previous findings that in these silver-based compounds silver is the major component that contributes to the broad spectrum of antimicrobial activities [28, 37]. Acticoat Antimicrobial Barrier Dressing and Sea Cell Active have similar inhibitory zone sizes regarding *S. aureus* while Sea Cell Active seems to be more effective against *E. coli.*

S. aureus is often responsible for the outbreak and aggravation of atopic dermatitis. *E. coli* is one of the most common bacteria in the genital and anal regions and responsible for infections in these areas.

Conclusions

The present results demonstrate the antifungal and antibacterial effect of Sea Cell Active in a standardized in vitro test system. Whether this fiber seems to be suited for bioactive textiles in specific anatomical body regions and skin conditions with a susceptibility to fungal and bacterial infections, namely *Candida* species, *S. aureus* and *E. coli,* must be examined by means of further investigations, especially human in vivo tests considering allergic and toxic effects of the fiber.

Acknowledgments

This study was supported by Alceru GmbH, Schwarza-Rudolstadt, Germany.

References

1 Altemus M, Rao B, Dhabhar FS, Ding W, Granstein RD: Stress-induced changes in skin barrier functions in healthy women. J Invest Dermatol 2001;117:309–317.
2 Ou-Yang H, Stamatas G, Saliou C, Kollias N: A chemiluminescence study of UVA-induced oxidative stress in human skin in vivo. J Invest Dermatol 2004;122:1020–1029.
3 Dhabhar FS: Acute stress enhances while chronic stress suppresses skin immunity: the role of stress hormones and leukocyte trafficking. Ann NY Acad Sci 2000;917:876–893.
4 Friedman EM, Lawrence DA: Environmental stress mediates changes in neuroimmunological interactions. Toxicol Sci 2002;67:4–10.
5 Elsner P: What textile engineers should know about the human skin. Curr Probl Dermatol 2003;31: 24–34.
6 Maurer D, Stingl G: Immune mechanisms of atopic dermatitis. Wien Klin Wochenschr 1993;105:635–640.
7 Fisher GJ, Kang S, Varani J, Bata-Csorgo Z, Wan Y, Datta S, Voorhees JJ: Mechanisms of photoaging and chronical skin aging. Arch Dermatol 2002;138:1462–1470.
8 Bolognia JL: Aging skin. Am J Med 1995;98:99–103.
9 Fenske NA, Lober CW: Skin changes of aging: pathological implications. Geriatrics 1990;45: 27–35.
10 Zikeli S: Sea Cell® Active – A new cellulosic fiber with antimicrobial properties. Avantex – International Forum and Symposium for High-Tech Apparel Textiles, Frankfurt/Main, 2002.
11 Zikeli S, Wolschner B, Eichinger D, Jurkovic R, Firgo H: Process for producing solutions of cellulose. Patent EP 0356419. 1990.
12 Bauer R, Taeger E, Zikeli S: Alceru – Manufacture of cellulosic materials by using organic solvent process. Seminar Cellulosic Man-Made Fibers in the New Millennium, Stenungsund, June 2000.
13 Zikeli S, Ecker F: Method for spinning a spinning solution and spinning head. Patent WO 01/81663. 2000.
14 Zikeli S, Ecker F: Verfahren und Vorrichtung zur Herstellung von Endlosformkörpern. Patent DE10037923. 2000.
15 Zikeli S: Lyocell fibers with health-promoting effect through incorporation of seaweed. Chem Fibers Int 2001;51:272–276.
16 Zikeli S: Lyocell-Fasern mit gesundheitsfördernder Wirkung durch Inkorporation von Algen. 7. Symposium Nachwachsende Rohstoffe für die Chemie, Dresden, 2001, vol 18, pp 449–466.
17 Zikeli S, Endl T, Martl MG: Polymer compositions and moulded bodies made therefrom. Patent WO01/62844. 2001.
18 Zikeli S, Endl T, Martl MG: Cellulose sharped body and method for the production thereof. Patent WO01/62845. 2001.
19 Strasburger E, Noll F, Schenck H, Schimper AFW: Lehrbuch der Botanik für Hochschulen, ed 33. Stuttgart, Fischer, 1991.
20 Hoppe HA, Levering T, Tanaka Y: Marine Algae in Pharmaceutical Science. Berlin, de Gruyter & Co, 1979.
21 Vasage M, Rolfsen W, Bohlin I: Sulpholipid composition and methods for treating skin disorders. Patent US 6,124,266. 2000.
22 Alban Muller International Product information – HPS3 – Padina Pavonica, 1999.
23 Hills CB: Extraction of anti-mutagenic pigments from algae and vegetables. Patent US 4,851,339. 1989.
24 Ruegg R: Extraction process for beta carotene. Patent US 4,439,629. 1984.
25 Henrikson R: Earth Food Spirulina, ed 5. Kenwood, Ronore, 1999.
26 Pedersen A: Studies on phenol content and heavy metal uptake in fucoids. Hydrobiologia 1984;116/117:498–504.
27 Leusch A, Holan ZR, Volesky B: Biosorption of heavy metals by chemically reinforced biomass of marine algae. J Chem Tech Biotechnol 1995;62:279–288.
28 Russel AD, Hugo WB: Antimicrobial activity and action of silver. Prog Med Chem 1994;31: 351–370.

29 Mundy GR, Guise TA: Hormonal control of calcium homeostasis. Clin Chem 1999;45:1347–1352.
30 Perianin A, Synderman R: Analysis of calcium homeostasis in activated human polymorphonuclear leukocytes: evidence for two distinct mechanisms for lowering cytosolic calcium. J Biol Chem 1989;264:1005–1009.
31 Millard PJ, Roth BL, Thi HP, Yue ST, Haugland RP: Development of the FUN-1 family of fluorescent probes for vacuole labeling and viability testing of yeasts. Appl Environ Microbiol 1997;63: 2897–2905.
32 Rodgers GL, Mortensen JE, Fisher MC, Long SS: In vitro susceptibility testing of topical antimicrobial agents in pediatric burn patients: comparison of two methods. J Burn Care Rehabil 1997;18:406–410.
33 Grier N: Silver and its compounds; in Block SS (ed): Antiseptics and Disinfectants. Philadelphia, Pa., Lea & Febiger, 1977, pp 375–389.
34 Johnson & Johnson: Product information – Actisorb Silver 220. 2001.
35 Tex-A-Med Product information – Padycare®. 2001.
36 Ward RS, Saffle JR: Topical agents in burn and wound care. Phys Ther 1995;9:547–559.
37 Rosenkranz HS, Rosenkranz S: Silver sulfadiazine: interaction with isolated desoxyribonucleic acid. Antimicrob Agents Chemother 1972;2:373–383.
38 Modak SM, Fox FL: Binding of silver sulfadiazine to the cellular components of *Pseudomonas aeruginosa*. Biochem Pharmacol 1973;22:2391–2404.
39 Bragg PD, Rainnie DJ: The effect of silver ions on the respiratory chain of *Escherichia coli*. Can J Microbiol 1974;20:883–889.
40 Woodward RL: Review of the bactericidal effectiveness of silver. J AWWA 1963;55:881–886.
41 Haynes L, Schulte TH: Antibacterial silver surfaces – An assessment of needs and opportunities for clinical devices. First International Conference on Gold and Silver in Medicine, Marylena, May 1987.
42 Yin HQ, Langford R, Burrell RE: Comparative evaluation of the antimicrobial activity of Acticoat antimicrobial barrier dressing. J Burn Care Rehabil 1999;20:195–200.

Dr. rer.nat. Uta-Christina Hipler
Department of Dermatology and Allergology
Friedrich Schiller University of Jena
Erfurter Strasse 35
DE–07740 Jena (Germany)
Tel. +49 03641 9 37355, Fax +49 03641 9 37437, E-Mail Christina.Hipler@med.uni-jena.de

Burg G (ed): Biofunctional Textiles and the Skin.
Curr Probl Dermatol. Basel, Karger, 2006, vol 33, pp 179–199

Antimicrobial-Finished Textile Three-Dimensional Structures

M. Heide[a], *U. Möhring*[a], *R. Hänsel*[b], *M. Stoll*[b], *U. Wollina*[c], *B. Heinig*[c]

[a]Textilforschungsinstitut Thüringen-Vogtland eV, Greiz, [b]Forschungsinstitut für Leder und Kunststoffbahnen gGmbH, Freiberg, und [c]Hautklinik am Klinikum Dresden-Friedrichstadt, Dresden-Friedrichstadt, Deutschland

Abstract

This paper describes the possibilities of antimicrobial finishing of three-dimensional spacer fabrics and its applications, and gives information about the different effects. A research project of the Textilforschungsinstitut Thüringen-Vogtland Greiz is presented in which medical shoe insoles, based on specially manufactured three-dimensional spacer fabrics, made of permanently effective antimicrobial yarns were used for interesting and functional textile products. Furthermore, work of the research institute Forschungsinstitut für Leder und Kunststoffbahnen Freiberg is presented which describes the silver-coating process and application of textile materials using antimicrobial substances. The chemical and mechanical stability is investigated, and proof of the effectiveness is supplied. The results show that in the three-dimensional spacer fabrics both – antimicrobial yarn materials and thin silver films with antimicrobial substances – can achieve an antimicrobial effect, even in low quantities.

Initial Situation

Micro-organisms are part of our daily life. They join humans in different forms – mostly unnoticed. Especially our skin is home for many of these creatures.

Micro-organisms can basically be divided into bacteria, mildew, yeast and viruses. Close contact with the skin, long periods between washings and a special microclimate can favour a fast growth of germs. These bacteria and fungi and their decomposition products cause

- infections,
- unpleasant odours in clothes, sportswear and shoes,
- allergies.

During the last few years, the materials for manufacturing textiles have shown positive tendencies towards a higher functionality. With fibres like X-Static®, Meryl Skinlife®, Diolen Care®, Trevira® Bioactive and others, the market was enriched with innovative antimicrobial products.

The following applications are known so far:

- technical textiles,
- working clothes,
- outdoor clothes,
- home clothes,
- sportswear and leisure wear,
- wellness,
- medical textiles.

Silver fibres or enclosed silver ions prevent the growth of bacteria without interrupting the natural balance of the skin. Odours, often caused by corynebacteria [1], are reduced. A longer, fresh feeling for the day is noticeable. The antimicrobial finishing of textiles must be deep-seated and permanent (fast), highly efficient against bacteria and fungi, non-toxic for the user and at the same time provide the usual wear comfort and function. The fibres mentioned above have already reached a remarkable market volume for clothes or functional clothing. Dermatological tests certified a good skin compatibility [2]. Thus, these textile products could not only find a domain in the wellness sector. In fact, the goal is to use textile fabrics with antimicrobial finishing sufficient for prophylaxis and therapy.

Preparation of Textiles with Antimicrobial Agents

The Use of Silver Ions on a Scientific Basis

Close co-operation of medics and textile technicians allows more and more the development of specific knitted fabrics which fulfil the demands of medical-clinical use. Besides features like breathability, moisture transport, antistatic properties, sterility, washability, minimization of allergies, cost-efficiency and ecological recycling, an antimicrobial protection with a barrier function against micro-organisms gets more and more important [3]. Therefore, a textile preparation with silver ions is preferable.

Silver has been known for centuries for its anti-infective quality and has been established in several of today's medical areas. At the beginning of the 20th century, silver and its chemical compounds were used as standard to fight bacterial infections [4]. Silver was used as thin film to heal wounds.

Different isotopes of silver are known. It can exist as a neutral element with 47 electrons and 47 protons or as a positively loaded atom with 46 electrons and 47 protons. A silver atom without electric charge is often mentioned as metallic silver or Ag (0) and has no antimicrobial properties. This feature is only available in the ionic mode Ag (I) or Ag^+. The silver cation is a potent antimicrobial agent to damage bacteria [5].

In vitro studies showed antibacterial effects and the creation of an antimicrobial barrier for silver-coated wound bandages over a period of several hours [4]. Schaller et al. [6] reported a good antibacterial effect by using silver-bearing wound bandages without induced cell alterations and cytotoxicity. With numerous animal experiments the use of silver in open wounds was investigated [7, 8].

The exact effect regarding the antimicrobial potential of Ag^+ has not been completely defined until today. The area where silver ions cause a dissociation of the cytoplasmic membrane from the cell wall has been described so far. The influences of protein interaction, enzyme activity [9] as well as DNA replication [10] need to be discussed. Silver toxicity was reported by different authors [7, 11]. The toxicity of silver-bearing medical products in humans can be rated as marginal for local or inner use [12].

In modern medicine, silver-bearing materials are used for some therapeutic products, for instance wound pads (silver-active coal compresses), special bandages (for burns) or catheter coatings under the aspect of anti-infective effects [12].

Good therapeutic experiences with silver-coated textiles are already available for the therapy of atopic eczema [2], where colonization by *Staphylococcus aureus* can be proved nearly every time.

For patients with diabetes mellitus, a complete therapy consists not only of an optimal adjustment of metabolism with medicamentous and dietetic measures, but also a continuous control of the infection. Especially with a sensorimotor or autonomous neuropathy, patients tend to a diabetic foot syndrome [13]. Consistent local pressure release, anti-infective care [14] and partly systemic antibiosis [15] are necessary to prevent after effects like amputation. The implementation of silver ions into shoes or socks for prophylaxis can be very helpful for diabetics to keep their mobility.

Several developments were made during the last few years by thread manufacturers to enwrap synthetic yarns with silver or to place silver ions into thread materials. Textile fabrics can be produced for medical products and incorporated into different final products, whether as a fabric or knitted fabric or fleece.

The positive properties of silver are e.g. [5, 12, 13]:

- antimicrobial properties,
- broad-band effects against bacteria,
- temperature-regulating properties,

- high heat conductivity, change of moisture (evaporation),
- enhancement of blood circulation,
- antistatic effect,
- odour control.

Negative features are comparatively rare. Silver can exert irritating and adstringent influences on the skin. With systemic resorption, argyrosis, CNS and nephrotoxicity [11] are possible. This regards especially organic silver applications. Ionic silver is well fabric compatible.

Antimicrobial Finished Textiles

Table 1 gives the specifications of antimicrobial finished textiles.

Application of Active Ingredients

Silver ions are used in medical textiles for the prophylaxis and therapy of diseases where micro-organisms have a pathogenetic effect.

The application of silver can occur by embedding the active ingredients during the spinning process or as a finishing process during the yarn or fabric production. Advantages and disadvantages are shown in table 2.

It is important to distinguish between:

- classical finishing of textiles with bio-active substances, where the active substance is on the surface and diffuses from the thread into the environment;
- processing of silver-coated synthetic yarns on the basis of polyamide (PA) 6.6 to textile surfaces, where the ratio of silver is normally approximately 15%;
- possible combinations are: combed cotton (Mako) + 5%, 7% or 10% X-Static or Shieldex®, polyethersulphone (PES) half matt + 5, 7 or 10% X-Static or Shieldex,
- PES T402 + 5, 7 or 10% X-Static or Shieldex.

Modified synthetic yarns are used on the basis of PA 6.6 or polyester. Active substances are silver ions as they are used in medicine or for drinking water treatment. To assure the stability of the silver ions they are put into zirconia-based ceramics and placed into the fibre during the spinning process. This guarantees permanent activity.

Antimicrobial Finishing of Textile Three-Dimensional Structures for Medical Insoles

Three-dimensional spacer fabrics, developed by the Textilforschungsinstitut Thüringen-Vogtland (TITV), are an interesting textile technological alternative for medical applications. They provide potential for problems where specific

Table 1. Specifications of antimicrobial textiles (selection)

Manufacturer	Product	Application	Specification	Permanence
Trade marks				
Odlo	Termic	sports underwear	silver ions on small ceramic layer	permanent
Tex-A-med	Padycare®	bedclothes, T-shirts, leggings	silver-coated microfibres PA 6.6, approx. 20% silver	150 washings at 40°C
Schoedel AG	Silverline®	cover fabric for mattresses	silver-coated fibre	permanent
Schoedel AG	Microcare®	cover fabric for mattresses	acetate with additive, which produces active O_2 with moisture, 'oxygen disinfection'	repeatedly washable up to 60°C
Malden Mills	Polartec® Power with Dry X-Static		implemented silver-coated threads	permanent
Antimicrobial fibres				
Acordis	Amicor™ Amicor Plus	sportswear, underwear, socks, medical products	PAN, only in combination with CV or CO, WO/PES, contains triclosan	
TWD GmbH	Diolen Care® Timbrelle Care®	sportswear, clothes, mattresses, medical products	PES or PA 6.6 multifilament with silver ions	permanent
Lenzing	Modal® Fresh	shirts, socks, clothes, bedclothes	viscose with antibacteriostatic activity as in toothpaste, cosmetics	more than 50 washings
Kanebo	Livefresh	underwear, socks, sportswear, soles	PA 6.6, often in combination, zeolite and silver ions	more than 50 washings
Montefibre	Leacril® Saniwear Terital® Saniwear	sportswear, bedclothes, tablecloths	PAN, PES, silver ions	
Statex Noble fibre	Shieldex® X-Static®	antineurodermatitis clothes, socks, underwear	silver-coated PA 6.6	permanent
Nylstar	Meryl® Skinlife	sportswear, socks, shoe linings, underwear, medicine	PA 6.6 microfibre with silver ions	permanent
Rhovyl	Rhovyl AS®	hygiene articles (incontinence, bandages)	PVC hollow fibre, triclosan	

Table 1. (continued)

Manufacturer	Product	Application	Specification	Permanence
Trevira GmbH	Trevira® Bioactive	functional underwear, sportswear, bedclothes	PES multifilament with fixed, antimicrobial additives	permanent

CO = Cotton; CV = viscose; PA = polyamide; PAN = acrylic; PES = polyethersulphone; PVC = polyvinyl chloride; WO = wool.

Table 2. Advantages and disadvantages of silver applications (source: Textil Color AG)

Application	Advantages	Disadvantages
After treatment	easy metallization, flexible costs	bactericidal impact, not permanent, migration into skin possible (allergenic), environmental influences
Silver coating	permanent, no migration	high costs (up to 10% Ag), achromatophil, metallic silver (discoloration), effect only given by contact with the skin
Antimicrobial yarns	antimicrobial, no migration, permanent, multifilament, no environmental influences	effect only given by contact with the skin and the release of silver ions

functions of the final product are required. Textile-physiological tests proved properties like pressure release, thermoregulation and moisture transport, and favoured spacer fabrics for the use in medicine [16].

Today typical applications can be already named, for instance primary prevention of pressure lesions caused by long-term operations or secondary prevention of decubitus for wheelchair drivers and the use of spacer fabrics as a therapeutic utility like bandages or orthoses.

The latest research of the TITV concerns the development of transthermal therapeutic systems like active ingredients in plasters with spacer fabrics as a depot for non-crystalline agents and the development of three-dimensional knitted compression bandages with defined elasticity for the complex anticongestion therapy in the treatment of lymph oedema.

Within a Zutech project (Zutech = Future Technologies for Small and Medium Enterprises) the partners TITV Greiz, Research Institute for Leather and Plastics gGmbH Freiberg and Test and Research Institute for Shoes e.V. Pirmasens research the development of functional spacer fabrics to prevent plantar ulcers for patients with diabetic foot syndrome. This application needs completely different construction parameters. Three-dimensional knitted fabrics will be used with traditional sole inlays for the orthopaedic shoe as well as a close-to-the-foot textile layer for working shoes. An advantage which should also be mentioned is the implementation of antimicrobial materials into the three-dimensional structure.

Problem

There are more than 4 million diabetics in Germany today [17] with the number increasing. Statistics estimate a number of approximately 1.25 million patients with the possibility of ulcerations and the risk of an amputation.

The diabetic foot syndrome is a long-term consequence. The feet and lower legs belong to the most sensitive body parts of a diabetic regarding the sensorimotor and autonomous neuropathy [13]. 40–70% of all non-traumatic amputations of the lower extremities are carried out on patients with diabetes mellitus [18]. Prostheses or individually manufactured orthoses are necessary.

In the therapy of diabetic foot syndrome, local pressure release of lesions takes priority and therefore the supply with dimensionally accurate shoes. Today, different materials for those shoes are used.

Orthopaedic inlays have the following functions: correction, support and bedding of foot deformations; release and redistribution of pressure in the lower extremities as a functional unit.

Besides cork, leather, metal and wood, mostly plastics are used. Herewith it is possible to adjust the function of the inlay correctly. Thermoplastic plastics are suitable for supporting and corrective inlays, soft plastics and silicone for highly sensitive feet. Harder foams are used for subconstruction or supporting inlays.

Textile materials have not played such an important role so far, they are used as cover layer (towelling, felt, fabric) for other materials like cork or foam. Thus, the development of functional three-dimensional structures as an alternative or useful supplement for inlay materials becomes more important.

Forcing of new product properties leads to the further development of existing products or to new ones with additional value. It is important to achieve new innovations for patients or occupational groups to increase their well-being.

Development Goals

The goals of development were functional three-dimensional textile structures with practical value-oriented construction and antimicrobial finishing achieving specific features for the use in shoes. The attention turned to the medical care of diabetics as well as to working shoes.

The goals are:

- consistent pressure release and soft bedding of feet to prevent lesions,
- establishing a permanent pressure-stable climate zone for ventilation even under the pressure of the human body,
- maintenance of form to avoid folding and pressure marks,
- use of functional materials or coatings with antimicrobial finishing for the prevention of injuries,
- tests and manufacture of prototypes.

These demands were fulfilled by the development of functional three-dimensional knitted fabrics in combination with approved foot bedding materials and suitable lamination processes.

The manufacturing technology is like no other textile process predestined to ensure a soft bedding of the feet and thermoregulation by the specific combination of materials and a suitable layout of the distance-keeping zone of the fabric. The rolling motion of the foot supports the circulation of air in the mentioned area.

Spacer fabrics are an ideal solution for the inner shoe due to their abilities to release pressure, improve climate circulation and their antimicrobial finishing. Three-dimensional fabrics with moisture-draining function are developed, tested and used in working shoes and for diabetics.

Specific Properties of Functional Spacer Fabrics

Spacer fabrics are three-dimensional manufactured knitted fabrics. They consist of two outer textile layers which are linked with pile threads. These pile threads provide a defined distance between the outer layers, which varies from 1.5 to 10 mm (fig. 1). The construction influences the functionality of the three-dimensional structure regarding thermoregulation, breathability, pressure stability and pressure elasticity. Both of the outer layers can be manufactured differently. Material and layout of the surface have an influence on elongation-elastic properties, the clothing-physiological comfort and the moisture transport away from the skin into the second layer of the textile structure. Spacer fabrics fulfil in multiple ways the expectations for textiles in medical use. The distance-keeping zone enables the circulation of air between the outer layers so that heat congestion and maceration of the skin will be avoided. Wearing comfort increases and leads to a better patient compliance for medical and therapeutic applications. Depending on the field of application the pressure-elastic behaviour of the structure may be altered, e.g.explicit pressure release in decubitus prophylaxis or in shoes

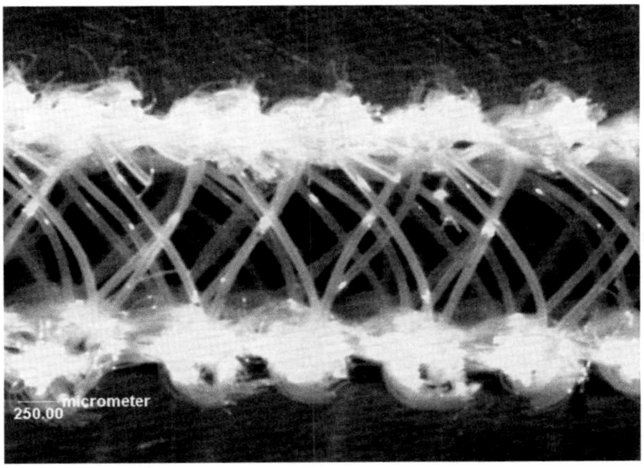

Fig. 1. Profile of a three-dimensional spacer fabric.

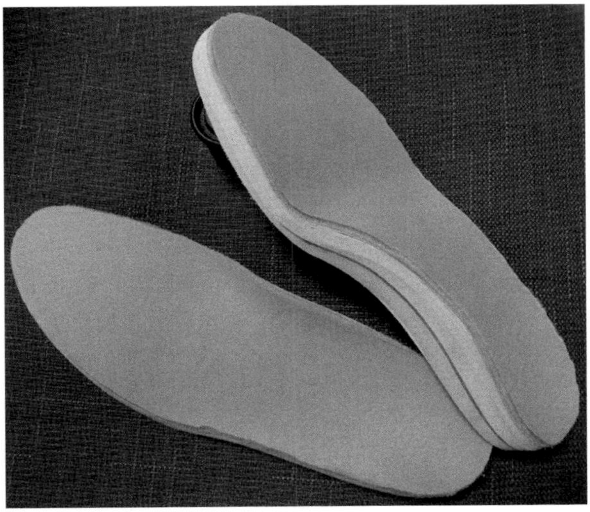

Fig. 2. Medical shoe inlay with antimicrobial-finished textile three-dimensional structure.

The Use of Modified Functional Synthetic Yarns for the Feet

On the basis of a coordinated test plan, a fabric for the use in orthopaedic or working shoes was developed. Different combinations of materials were tested, and their influence on pressure release and antimicrobial effect was investigated (fig. 2).

Table 3. Properties of synthetic and cellulosic yarns

Fibre	Fibre type	Density, g/cm²	Tear lengthening, %		Electrical resistance, Ω/cm	Melting temperture, °C	Water sorption, mass %	Water retention %
			climate	wet				
Synthetic fibres								
PA 6	filaments	1.14	20–45	105–125	10^9–10^{11}	215–220	3.5–4.5	10–15
PA 6.6	filaments	1,14	20–40	105–125	10^9–10^{11}	255–260	3.5–4.5	10–15
Polyester	filaments + strands	1.36–1.41	20–30	100–105	10^{11}–10^{14}	250–260	0.2–0.5	3–5
Cellulosic fibres								
Viscose	filaments	1.52	10–30	100–130	low (10^6–10^7)	175–190	12–14	85–120
Cotton	fibres	1.53–1.55	20–50	100–120	low	above 180	7–9.5 at 65% humidity 14–18 at 95% humidity	42–53

Materials

The functionality of spacer fabrics depends on the used materials. Often, synthetic yarns like PA 6.6 or polyester are used but celluloid yarns like cotton or viscose too. Table 3 gives a review of the materials.

Synthetic Strands. For distance-keeping, mostly synthetic strands spun by single threads are used. They provide higher stiffness than a multifilament of the same thickness and have a better pressure stability. Strand yarns are not able to keep and forward moisture, which makes them different from filament yarns.

Synthetic Filament Yarns. Filament yarns are made by pressing the spinning mass through nozzles with different openings. This creates a certain number of thin threads (filaments) which are combined to a single thread. The hollow space between the single filaments can transport moisture by adhesive force. The more filaments there are, the better the moisture transport. Spacer fabrics, for instance, can drain sweat away from the skin.

Synthetic Multifilament Yarns with Antimicrobial Finishing. An additional value of manufacturing synthetic yarns can be the implementation of silver ions and herewith an antimicrobial effect. Nanocrystalline silver is incorporated into the spinning process. A permanent antiseptic effect is given.

The positive silver ions inhibit the function of bacterial enzymes. They attack the structure proteins of the germs and inhibit cell division. This leads to denaturation. For the antimicrobial finishing of the zone which is close to the foot, newly developed synthetic yarns on the the basis of PES (Diolen Care or Trevira Bioactive) and PA 6.6 (Meryl Skinlife) were used.

A selective effect on lesions which can have disastrous consequences for diabetics was aimed at. Spacer fabrics No. 09960/3 and 12980/5 were modified with antimicrobial synthetic yarns just as No. 02991/2 for the use in working shoes. Besides the antimicrobial effect, the protection against mycosis and odour was important here.

Test for Antimicrobial Effect

You will always find bacteria on a healthy skin. This colonization protects our organism. There are Gram-positive micro-organisms, like staphylococci and several corynebacteria, and Gram-negative bacteria, like *Klebsiella pneumoniae* or *Escherichia coli* [19]. With the presence of moisture, bacteria increase their number by cell division rapidly with or without oxygen. The antimicrobial effect was tested for both, Gram-positive (*Staphylococcus aureus*) and Gram-negative (*K. pneumoniae*) microbial strains referring to the challenge test, test method JIS L 1902-1998 [20]. This test includes textiles with antimicrobial finishing as well as textiles with immobilized agents.

With the given conditions for spacer fabric No. 12980/A (finished with PES Diolen Care), an antimicrobial effect was proved against *S. aureus* and *K. pneumoniae* DSM 789. For spacer fabric No. 09960/3 with PA 6.6 Meryl Skinlife, an antimicrobial effect was proved against *K. pneumoniae* DSM 789 und *S. aureus* ATCC 6538 in comparison with the non-finished sample.

Figure 3 shows the test results for spacer fabric No. 09960/3.

The first and second pairs of columns show the reference standard of 0 h (without antimicrobial-finished materials) and the number of colonies after 18 h of incubation. The third pair shows the colony number of the antimicrobial-finished fabric after 18 h and that on the right the specific activity.

Further Options for Antimicrobial Finishing of Three-Dimensional Structures

Silver-Coated PA Threads. Silver-coated PA threads from Statex Co., Bremen, named Shieldex are known as Padycare® clothes of Tex-A-Med Co., Gefrees, for neurodermatitis patients. The silver-covered knitted fabrics can be processed to textile surfaces. Microfibres with PA 6.6 are coated with silver which is locked in the surface of the microfibre (fig. 4).

To work single threads into spacer fabrics is possible, but too expensive.

Use of Silver-Coated Warp Knittings. During the development a knitted fabric, Shieldex of Statex Co., was laminated on spacer fabrics and tested in the

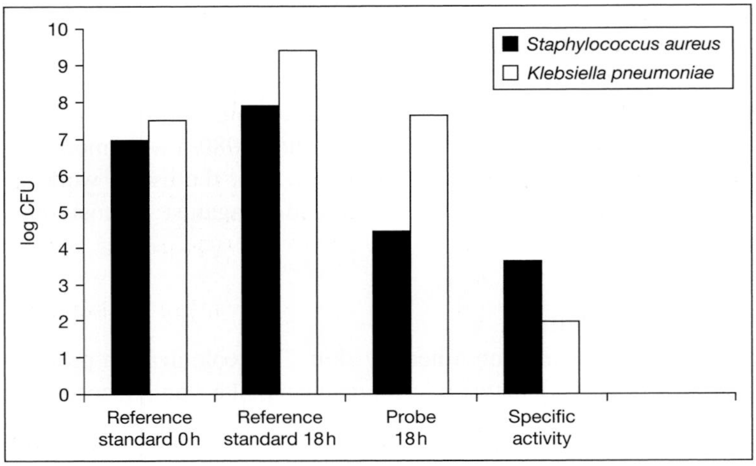

Fig. 3. Proof of antimicrobial effect of silver ions with test method JIS L 1902 for spacer fabric 09960/3. Specific activity = reference at 18 h – probe at 18 h.

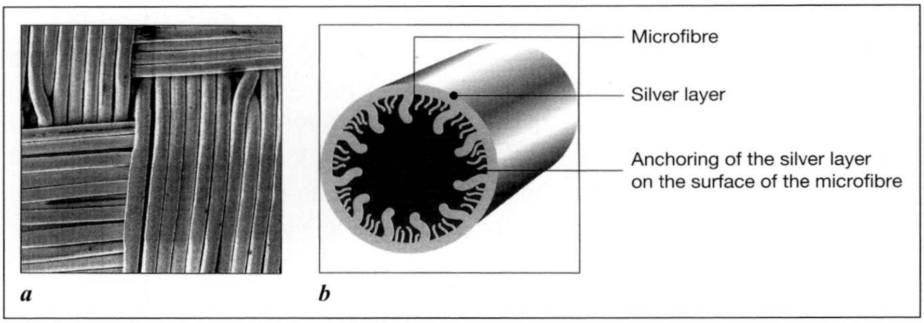

Fig. 4. Silver-coated PA fibres (source: Tex-A-Med GmbH). ***a*** Microphotograph of the silver-coated microfibres. ***b*** Cross-section of the microfibre. Even frequent washing does not detach the silver layer.

area close to the foot. The positive properties of spacer fabrics originate from the precise structure of the silver-covered knitted fabric.

Antimicrobial Finishing of Three-Dimensional Warp Knitting

The demand for antimicrobial-finished textiles is increasing. Besides functionality, it is necessary to achieve skin and environmental compatibilities. The task is to add antimicrobial finishing durably onto the textile fibre in order to get products with a permanent effect.

Options to Prevent Microbial Contamination

There are three trends:

- antimicrobial finishing which prevents accumulation of micro-organisms (bacteriophobe finishing),
- antimicrobial finishing which inhibits the growth of accumulated micro-organisms (bacteriostatic finishing),
- antimicrobial finishing which kills adhering micro-organisms (bactericidal finishing).

Tests

The Research Institute for Leather and Plastics made investigations to give three-dimensional knitted fabrics an antimicrobial finishing afterwards. The test was made with their own equipment using technologies like the plasma process and the magnetron sputter process. Both technologies allow to coat nanolayers onto fibres, but keeping the textile properties. The plasma functionalizes the textile fibres during separation. That means that radical spots are made in the polymer matrix where functional groups (hydroxyl, keto or carboxyl) assure a strong anchorage of the antimicrobial finishing.

Silver Coating to Prevent Accumulation of Bacteria

Silver inhibits the growth of bacteria. The advantage of silver ions is their effect on the environment and that they do not need to be dispersed more widely [21]. With the nanothin layers which can be coated with the help of this technology, a contribution to cost-effectiveness is made, too.

Application of Silver Layers with the Sputter Process
(Separation of Plasma)

The coating of three-dimensional knitted fabrics with a silver layer was made in a laboratory. The sputter process differs from the evaporating process in dispersing the metal from the solid condition and not after conveying it into the gaseous condition.

In this case, a target (silver) and the substrate (three-dimensional knitted fabric) are positioned a few centimetres away from each other. Between these two surfaces (which act as electrodes), a plasma discharge takes place within an argon atmosphere. The resulting argon ions and electrons move at high speed to the target and release atoms or its fragments from the surface. These parts then move to the substrate surface. In comparison with the thermal vaporization, the

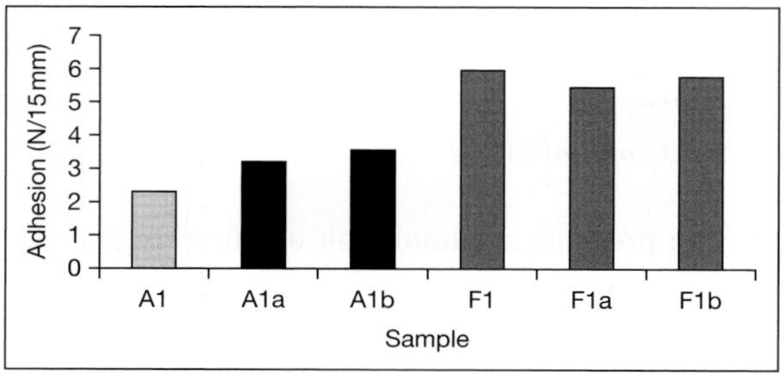

Fig. 5. Adhesion of silver layers to different surfaces. Fabric 12980/5: A1 = no pretreatment, 100-nm silver layer; A1a = oxygen-plasma pretreatment, 100-nm silver layer; A1b = as A1a, additional protection layer SiO_x; fabric 02991/2: F1 = no pretreatment, 100-nm silver layer; F1a = oxygen-plasma pretreatment, 100-nm silver layer; F1b = as F1a, additional protection layer SiO_x.

atoms have a 100–1,000 times higher kinetic energy which makes the layer more durable on the textile surface.

Test of Mechanical Properties of the Separated Layers. The silver layers were applied to different fabric structures. Spacer fabric No. 12980/5 consists of 95% polyester and 5% spandex, spacer fabric No. 02991/2 is 100% polyester. The spandex has a negative influence on the silver layer. To improve its processing properties, it is prepared with silicone.

The silver coating was made on untreated three-dimensional knitted fabrics and on surfaces pretreated with plasma. In another case, the silver layer was supplied with a protection layer on the basis of SiO_x. The coating process was made with plasma polymerization.

The mechanical stability was tested with adhesive tape to determine adhesion (peel test) and the abrasion resistance with the friction test (Martindale) [22] and friction fastness (Veslic) [23].

Figure 5 shows the results of the adhesion test. It can be seen that adhesion can be improved by additional plasma treatment with oxygen. A further improvement can be achieved by hydrophobic protection layers.

In addition, there is no need for plasma treatment if the fabric contains no spandex.

Because the spacer fabrics will be used as sole material for shoes, the abrasion test is more significant than the adhesion test. In this test, the stress during walking is simulated.

In the Veslic test, cotton as friction element was rated on a grey scale as well as the friction line on the test object. The results are shown in figure 6. After 3,000 friction runs, silver marks on the friction elements were visible. In contrast to the

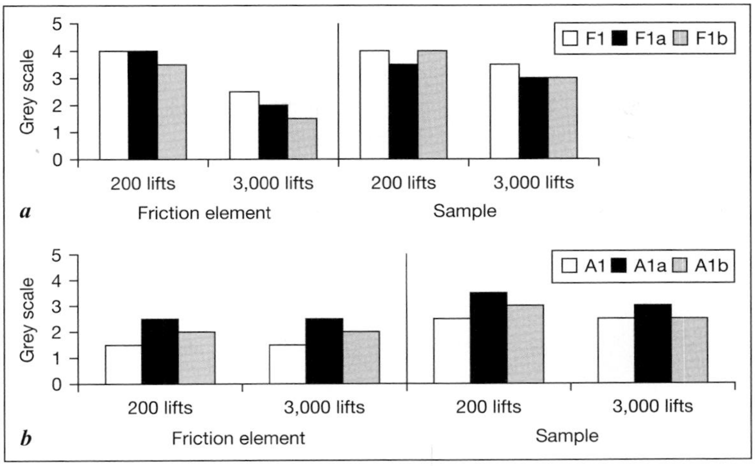

Fig. 6. Friction strain with cotton (200 and 3,000 friction runs). Rating of the friction element and the three-dimensional fabric surface was done with a grey scale: 5 = no abrasion, no visible marks on the friction line; 1 = high abrasion, no silver layer left on the fabric's surface. For fabric numbering, see legend of figure 5. ***a*** Abrasion of the silver layer on fabric 02991/Z. ***b*** Abrasion of the silver layer on fabric 12980/5.

adhesion test, the silver layer of article No. 12980/5 has a higher load capacity. A silver layer with 100 nm thickness should be preferred to one with 15 nm.

The Martindale test showed no significant change of the silver layer up to 10,000 rubbing runs of spacer fabric No. 12980/2. This fabric has a very homogeneous and close surface.

Article No. 02991/2 had a destroyed mesh structure after 10,000 runs (see the REM shots in fig. 7).

With energy-dispersive X-ray inclusion, the concentration of silver was rated before and after the rubbing tests on layers of 15 and 100 nm thickness. The results confirmed the expectations that the concentration of silver was higher in layers with 100 nm thickness.

The chart of the results is shown in figure 8.

Proof of Permanent Effect of the Layers

Washability. The test of resistance to washing was made with DIN EN 6330. The coating was not influenced by the washing process. The antimicrobial effect was improved. This can be caused by the cleaning process of the surface. An efficient cleaning of the spacer fabrics is recommended before the application of the silver layers.

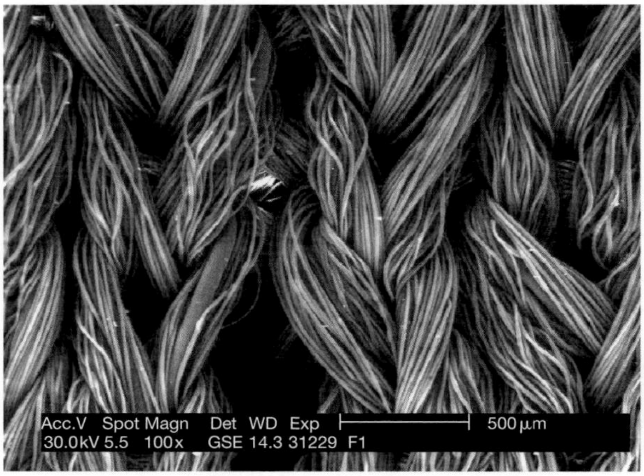

Fig. 7. Fabric 02991/2 after 10,000 friction runs.

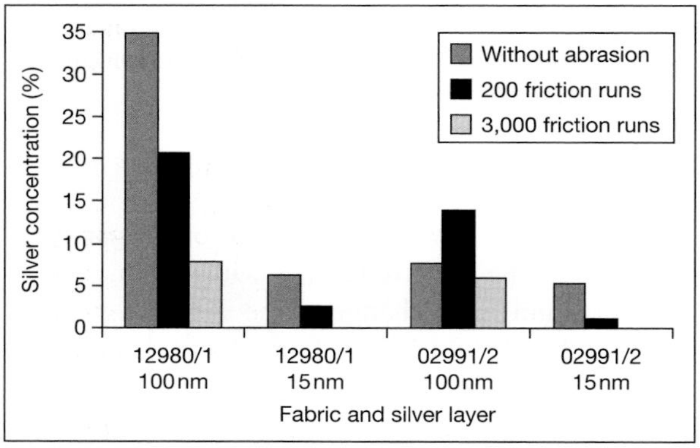

Fig. 8. Concentration of silver on the textile surface after friction strain.

Test of Antimicrobial Effect. The antimicrobial effect was tested by the IDUS (Biologische Analytische Umweltlabor GmbH) Company. Two test methods were used. Test method AA TCC TM-147 is the inoculation with *S. aureus* and *K. pneumoniae* and a rating of fouling after 24 h. This is a fast method for the determination of bactericidal effects and inhibition areolas. A rating for the decrease in bacteria on the textile surface through its antimicrobial finishing is not given. It only states whether the finishing is bactericidal or not.

With selected samples the agar diffusion bioassay method was used [24]. The test sample lies on a thin layer of culture medium for 24 h. This shows whether the finished textile surface can prevent the growth of bacteria in the

Table 4. Results of antimicrobial effect

Coating	*K. pneumoniae*		*S. aureus*	
	inhibiting areola	growth beneath the fabric	inhibiting areola	growth beneath the fabric
15-nm Ag layer	no	no	no	no
35-nm Ag layer	no	no	no	no
100-nm Ag layer	no	yes	no	no
100-nm Ag layer with SiO_x-	no	yes	no	no

contact area between textile and culture medium or not. If the finishing has no antimicrobial effect, a compact bacterial film grows.

The next step is the comparison of finished and non-finished surface.

The samples were inoculated with *S. aureus* (Gram-positive bacteria) and *K. pneumoniae* (Gram-negative bacteria).

Two different spacer fabrics were tested (polyester surface and a combination of polyester and spandex). The materials were coated with 15-, 35- and 100-nm silver layers. The 100-nm layer was protected with a plasma polymer layer on the basis of SiO_x.

The results are shown in table 4. The silver layers did not create inhibition areolas. *Klebsiella* showed a low sensitivity to thin silver layers. *Staphylococcus* reacted more sensitively to silver ions. All finishing samples have a bactericidal effect. With the agar diffusion bioassay, the antimicrobial effect of the silver ions on the surface after a friction test can be rated. The concentration of silver on the surface after the friction test was determined by energy-dispersive X-ray inclusion. Furthermore the biocidal effects of the different concentrations of silver were rated (table 5). The results show that the silver ions can decrease the growth of bacteria even after friction strain on the surface.

Separation of Organic Agents by Nanocoating with Atmosphere Pressure Plasma and Spray Coating

Products

The applied antibacterial agents were the following commercially available products:

Afrotin LC (Schill + Seilacher). This product is an aqueous solution on the basis of sodium pyridine dinthole. Danger: the product must not be introduced into the canalization or water. Insanitary effects are not known.

Afrotin FG (Schill + Seilacher). The basis is methyl-2-benzimidazole carbamate. Danger: the product must not be introduced into the canalization or water. Insanitary effects are not known.

Table 5. Effect of the concentrations of silver after friction test

	Concentration of silver on the surface, %	*K. pneumoniae*	*S. aureus*
12980/5 with 100-nm Ag layer without friction runs	34.79	no inhibiting areola, growth decreases to 50%	no inhibiting areola, nearly no growth visible
12980/5 with 100-nm Ag layer after 200 friction runs	20.62	no inhibiting areola, growth decreases to 50%	no inhibiting areola, growth decreases to 50%
12980/5 with 100-nm Ag layer after 3,000 friction runs	7.84	no inhibiting areola, growth decreases to 50%	no inhibiting areola, growth decreases to 50%
02991/Z with 100-nm Ag layer without friction runs	7.75	no inhibiting areola, growth decreases to 50%	no inhibiting areola, growth decreases to 50%
02991/Z with 100-nm Ag layer after 200 friction runs	13.9	no inhibiting areola, growth decreases to 50%	no inhibiting areola, growth decreases to 50%
02991/2 with 100-nm Ag layer after 3,000 friction runs	5.85	no inhibiting areola, growth decreases to 50%	no inhibiting areola, growth decreases to 50%

Fungitex ROP (Pfersee Chemie). The product is a non-ionic solvent-containing emulsion of fatty acid ester, an aromatic compound and a benzimidazole derivate. Insanitary effects are not known.

Parmetol DF 18 (Schülke & Mayr). The agent is a combination of different structured n,s-heterocyclic compounds which have a synergetic effect. Danger: the product must not be introduced into the canalization or water. It is biodegradable. Insanitary effects are not known.

Tinosan NW 200 (Ciba Speciality Chemicals). The product is a dilutable emulsion on the basis of 5-chloro-2-(2,4-dichlorophenoxy)phenol. Danger: the product must not be introduced into the canalization or water. Insanitary effects are not known.

Application of Antibacterial Agents

Description of the Processes. The application was carried out with two different processes. For the application of nanolayers, the atmosphere-pressure-plasma process was used. For thicker layers, spray coating was the preferred method.

The principle of the plasma process is the addition of aerosol by barrier discharge. This combines high-energetic and chemical modifying.

Table 6. Applied quantities

Product	Quantity, g/m²
Tinosan NW 200	1.8
	3.4
	6.3
Afrotin FG	2.6
	10
Afrotin LC	2.6
	10
Parmetol DF 18	1.5
	8.9
	13
Fungitex Rop	3
	7.4

Two electrodes are placed in a reaction chamber where the discharging process takes place. Between those electrodes the plasma is added with a micro-atomizer, which creates drops of less than 2 μm. The water vaporizes and this allows the application of nanolayers to the material.

A carrier gas is responsible for the transport of the exhalation drops, i.e. air. Non-used aerosol will be disposed by an extraction system. The aerosol can be dosed into the barrier discharge or right after that.

A corona process adjusts the application into the textile surface. Additionally, ultraviolet radiation results from the discharging process. Both lead to the modification of the surface by creating new functional groups like hydroxyl, keto or carboxyl groups. This helps the antibacterial agents to compound with the surface. The application quantities ranged from 1 to 3 g/m^2.

Before an application with a spray gun, the textile surfaces were modified with corona or gas phase fluorination to improve adhesion. The quantities applied with a spray gun were 10–15% higher.

Table 6 shows the products and the applied quantities.

Test of Antibacterial Effect. Test method AA TCC TM-147 is the inoculation with *S. aureus* and *K. pneumoniae* and a rating of fouling after 24 h.

With *Staphylococcus*, inhibition areolas were formed, depending on the applied quantity and the product.

The most explicit effect showed the product Afrotin LC which formed large inhibition areolas with the bacterium *Klebsiella*, too.

Furthermore the results showed that low quantities had a bactericidal effect on the three-dimensional knitted fabrics.

Table 7 shows the relations between the inhibition areolas and the quantity of the applied agent.

Table 7. Inhibiting areolas in relation to the quantity of the applied agent

Product	Quantity, g/m^2	Size of the inhibiting areola, mm	
		K. pneumoniae	*S. aureus*
Tinosan NW 200	1.3	no inhibiting areola	10
	3.4	no inhibiting areola	13
	6.3	8	15
Afrotin FG	2.6	no inhibiting areola	7
	10	no inhibiting areola	9
Afrotin LC	2.8	>30	17
	10.3	>30	18
	20	>30	>30
Parmetol DF 18	1.5	no inhibiting areola	5
	8.9	no inhibiting areola	9
	13	no inhibiting areola	10
Fungitex ROP	3	no inhibiting areola	8
	7.4	no inhibiting areola	10

Conclusion

The test results on the basis of three-dimensional spacer fabrics showed that with antimicrobial-finished thread materials made of PA or polyester as well as with thin silver layers and antibacterial agents in low quantities an antibacterial effect for these three-dimensional knitted fabrics can be achieved. Concerning the application of agents, a silver coating is more suitable because the bacteria of the skin flora are more protected. Silver-coated material or threads giving off silver ions are recommended because their effect is permanent and dermatologically harmless (fig. 3).

For the health care of diabetics, medical shoes with an antimicrobial finishing of the used three-dimensional textiles are important to prevent wound infections which can have fatal consequences for this group of patients. In most cases, long-term therapies and high medical costs can be expected.

References

1 Natsch A, Gfeller H, Gyrax P, Schmid J: Isolation of a bacterial enzyme releasing axillary malodour and its use as a screening target for novel deodorant formulations. Int J Cosmet Sci 2005;27:115–122.

2 Wulf A, Moll I: Silberbeschichtete Textilien – eine ergänzende Therapie bei dermatologischen Erkrankungen. Akt Dermatol 2004;30:28–29.

3 Wollina U, Heide M, Müller-Litz W: Stellen Medizin und Gesundheitswesen neue Anforderungen an die Textilqualität? Melliand Textilber 1998;79:552–555.
4 Holder IA, Durkee P, Supp AP, Boyce ST: Assessment of silver-coated barrier dressing for potential use with skin grafts on excised burns. Burns 2003;29:445–448.
5 Ovington LG: The truth about silver. Ostomy Wound Manage 2004;50:1–10.
6 Schaller M, Laude J, Bodewaldt H, Hamm G, Korting HC: Toxicity and antimicrobial activity of a hydrocolloid dressing containing silver particles in an ex vivo model of cutaneous infection. Skin Pharmacol Physiol 2004;17:31–36.
7 Lee AR, Moon HK: Effect of topically applied silver sulfadiazine on fibroblast cell proliferation and biochemical properties of the wound. Arch Pharm Res 2003;26:855–860.
8 Lansdown AB, Sampson B, Laupattarakasem P, Vuttivirojana A: Silver aids healing in the sterile skin wound: experimental studies in the laboratory rat. Br J Dermatol 1997;137:728–735.
9 Williams R, Doherty P, Vince D, Grashoff G, Williams D: The biocompatibility of silver. Crit Rev Biocompatibility 1989;5:221–243.
10 Feng QL, Wu J, Chem GQ, Cui FZ, Kim TN, Kim JQ: A mechanistic study of the antibacterial effect of silver ions on *Escherichia coli* and *Staphylococcus aureus*. J Biomed Mater Res 2000;52:662–668.
11 Lansdown AB: Silver. 2. Toxicity in mammals and how its products aid wound repair. J Wound Care 2002;11:173–177.
12 Silver S: Bacterial silver resistance: molecular biology and uses and misuses of silver compounds. FEMS Microbiol Res 2003;27:341–353.
13 Haslbeck M, Renner R, Berkau HD: Das diabetische Fusssyndrom. München, Urban & Vogel Medien und Medizin, 2003.
14 Wollina U: Der diabetische Fuss – eine Übersicht für Dermatologen. Z Hautkr 1999;74:265–270.
15 Hilton JR, Williams DT, Beuker B, Miller DR, Harding KG: Wound dressings in diabetic foot disease. Clin Infect Dis 2002;39(suppl 2):S100–S103.
16 Wollina U, Heide M, Swerev M, Billia M, Möhring U: Abstandsgewirke und andere Abstandsgewebe. Akt Dermatol 2004;30:8–10.
17 Tautenhahn J: Diabetische Ulcerationen. Hartmann Wundforum 1998;4:10–17.
18 Morbach S, Müller E, Reike H, Risse A, Spraul M: Diagnostik, Therapie, Verlaufskontrolle und Prävention des diabetischen Fusssyndroms. Diabetes Stoffwechsel 2004;13:9–10.
19 Burghardt F (ed): Mikroskopische Diagnostik, Stuttgart, Thieme, 1992, p 681.
20 Japanese Industrial Standard, JIS L 1902: Testing method for antibacterial activity of textiles. 1998.
21 Vohrer U, Trick I: Strategien der antimikrobiellen Ausrüstung von Bekleidungstextilien. Avantex-Symposium, Frankfurt/Main, November 2000.
22 DIN EN ISO 12947/2.
23 DIN EN ISO 11640: Leder-Farbechtheitsprüfungen – Bestimmung der Reibechtheit von Färbungen. 1998.
24 DIN EN ISO 20645: Textile Flächengebilde – Prüfung der antibakteriellen Wirkung – Agarplattendiffusionstest. 2004.

M. Heide
Textilforschungsinstitut Thüringen-Vogtland eV Greiz
Zeulenrodaer Strasse 42
DE–07973 Greiz (Germany)
Tel. +49 03661 611 315, Fax +49 03661 611 222, E-Mail m.heide@titv-greiz.de

Author Index

Subject Index